Interspecies Communication

Interspecies Communication

SOUND AND MUSIC BEYOND HUMANITY

Gavin Steingo

The University of Chicago Press CHICAGO AND LONDON

The University of Chicago Press, Chicago 60637
The University of Chicago Press, Ltd., London

Published 2024
Printed in the United States of America

33 32 31 30 29 28 27 26 25 24 1 2 3 4 5

ISBN-13: 978-0-226-83133-6 (cloth)
ISBN-13: 978-0-226-83136-7 (paper)
ISBN-13: 978-0-226-83135-0 (e-book)
DOI: https://doi.org/10.7208/chicago/9780226831350.001.0001

This publication is made possible in part by the Barr Ferree Foundation Fund for Publications, Department of Art and Archaeology, Princeton University.

Library of Congress Cataloging-in-Publication Data

Names: Steingo, Gavin, author.
Title: Interspecies communication : sound and music beyond humanity / Gavin Steingo.
Description: Chicago ; London : The University of Chicago Press, 2024. | Includes bibliographical references and index.
Identifiers: LCCN 2023026930 | ISBN 9780226831336 (cloth) | ISBN 9780226831367 (paperback) | ISBN 9780226831350 (e-book)
Subjects: LCSH: Human-animal communication. | Extraterrestrial beings. | Musicians, Black. | Humanism.
Classification: LCC QL776.S744 2024 | DDC 591.59—dc23 /eng/20230705
LC record available at https://lccn.loc.gov/2023026930

♾ This paper meets the requirements of ANSI/NISO Z39.48-1992 (Permanence of Paper).

To Helen Hwayeon
my everything person

To Helen Hwayeon
my everything person

Contents

Preface

Interspecies Communication is about human attempts to communicate with non-humans. It focuses on how those attempts are mediated by sonic expression, including (but not limited to) language and music. Because I am a musicologist by training, questions surrounding music and aesthetics figure prominently throughout. Indeed, this project was inspired by cases (fictional and real) of humans playing music to—and, ostensibly, with—animals.

The book focuses on humans who, in their endeavor to reach other beings, have gone beyond their immediate surroundings, exploring the ocean's depths and the outer reaches of space. In this book's pages, the reader will encounter a motley group of individuals (scientists, philosophers, artists, musicians), as well as a diverse assortment of what Michael Schetsche and his collaborators call "maximal strangers."[1] In my own account, maximal strangers include domesticated animals, birds, dolphins, and whales, as well as extraterrestrials.

The book is anchored by two main arguments. The arguments may seem contradictory, but they actually work together. The first argument is that humans do, in fact, communicate with non-humans in numerous ways. That my dog heeds my call is no figment of my imagination. Dolphins can, indeed, learn a complex repertoire of instructions from humans. Human/animal communication is not merely a social construction, not merely a language game with nothing on the other side of it. As I argue in the pages that follow, adequately capturing interaction on this register requires developing a model of communication prior to, and outside of, language. Humans communicate with animals all the time, but we do not speak with them, since they do not possess the capacity for language. Other arrangements are in operation, other systems are at play.

The second argument is that, although we can and do communicate with non-human animals, this communication—which is often fairly rudimentary, happening at the level of instruction, for example—also gives rise to less obvious, less rational, and less scientific feelings and ideas. For example, dog owners may teach their pets to sit or roll over, but they also, frequently, say things like, "Good boy!" or "Come here you little munchkin, you know I love you!" Elsewhere, an ornithologist may study the lung capacity of a bird and sonic profile of a birdcall (and may admit that such calls have meaning), while a wanderer in the forest, or even the ornithologist herself, may wax lyrical about the beauty of birdsong—a beauty, many have felt, that mirrors the grandeur of Nature itself.

One might counter that emotions like love and aesthetic categories like beauty are themselves susceptible to scientific analysis, but my argument is that, when they function in the ways just described, they are not meant to be. They are meant, by contrast, to exceed the simpler forms of communication we have with the beings around us. In short, we communicate with animals, but we also desire them, fear them, and love them. If some of what we experience during our interactions with animals is hardly a figment of our imagination, then other aspects certainly are, and other aspects might be. Uncertainty, conjecture, and speculation are just as central to our experience as certainty is. I would even argue that when we approach so-called maximal strangers, humans have a tendency to transgress the rational and the reasonable. Immanuel Kant, whose entire philosophy was based on delimiting the boundaries of knowledge, observed an inevitable tendency to reach beyond what we can ever possibly know, to slide perpetually into delusion.[2]

Concrete interaction on one side; excess and flights of fancy on the other. Things get complicated when we put the two together. The third argument of this book (if there is one) is that the two registers go together—always. In other words, although they operate differently and require different forms of analysis, concrete interaction and its various excesses cannot be prized apart. The relationship is one of supplementarity, where the supplement is understood as something that both adds to and completes something else, thereby becoming part of it.[3]

In brief, this book asks not whether we communicate with non-humans, but how our attempts at communicating with non-humans force us to rethink the meaning of communication itself.

John Durham Peters's famous book on the history of communication helps clarify my argument. Peters is leery of the dialogic model of communication—that is, communication understood as the reciprocal sharing of ideas or interior emotions. But the impossibility of "shared minds," he says, does not mean that we "cannot cooperate splendidly."[4] In his view, cooperation trumps dialogic communication, which suggests that, "Whatever 'communication' might mean, it is more fundamentally a political and ethical problem than a semantic one."[5] While he tries to keep both registers at play (the politico-ethical and the semantic), he ultimately comes down on the side of the former. He shares Hegel's basic idea that "'communication' will always be more than the shuttling of mind-stuff. It is the founding of a world. For Hegel communication is not a psychological task of putting two minds *en rapport* but a political and historical problem of establishing conditions under which the mutual recognition of self-conscious individuals is possible."[6]

Grounding his theory of communication in ethics places love and desire (sexual and asexual) at the center of his inquiry. Peters critiques semantic notions of communication in order to develop a theory based on ethics, love, and world-building. There is a vector, in other words, from the semantic to the ethical, and that vector goes in only one direction. This way of thinking is evident in his brief analysis of interspecies communication in the sixth chapter of his book. There, he writes, "If the question were simply whether human and animals can cooperate in a variety of tasks, it would lack urgency; the answer would be obvious. The absence of 'communication' has never prevented humans and animals from entering into community with each other. Thousands of years of brute exploitation, and a shorter history of household pets, offer two alternative models besides 'communication.'"[7]

I like what Peters is doing here, and I mostly agree with him. But where he wants to shift focus from semantics to ethics, I want to keep both on the table. While he is happy to jettison the question of interspecies communication (he speaks of an *absence* of communication between human and animal), I want to think the question of communication and ethics together. Which is another way of saying that I want to think science and ethics together, as well as science and politics. Rather than shifting from the semantic-scientific to the ethical-political, this book is an attempt to dialectically mediate the two sides.

Whether I achieved my goal is up to the reader to decide. Obviously, I do not pretend to have solved the problem or to have settled the issue for good. But I hope that future discussions of interspecies communication attempt, as I have here, to maintain a balance between the semantic-scientific and ethical-political, rather than asserting the primacy of either one.

Those are the book's main arguments, but a book is more (although hopefully not less) than what it argues. When it comes to the topic of interspecies communication, and when one is faced with the various excesses I've outlined, some themes pop up so frequently that no writer can ignore them. The chapters of this book deal with these themes, and at times the arguments recede into the background. In this book, at least, arguments are not destinations; they are more like anchors, or perhaps guiding threads.

Interspecies Communication is divided into two parts, with two chapters in each. Part 1 deals with animals, and part 2 deals with life somewhere other than Earth—that is to say, with aliens. This may seem like a strange move (why give so much real estate to aliens, when they may not even exist?), but I've structured the book in this way because much thinking in this area is structured that way. From the marine scientist John Lilly to the philosopher Vilém Flusser and the technologist Jaron Lanier, people often turn to the extraterrestrial when trying to understand the weirdness of animal life. What is more, especially between the 1950s and 1970s, the same scientists used the same technologies for scanning the ocean's depths and for exploring outer space. The heyday of marine mammal science (when many important discoveries about dolphin and whale behavior were made) coincided with the main period of the Space Race (1955–1975).

In this sense, the choices made in this book are not my own. I document personal, social, and material linkages in the historical record. The same goes for other themes I address in the pages that follow. A reader uninitiated in the topic may be surprised by it, but anyone who has read around in these strange regions will be aware of the importance of music, love, and spirituality to many of the most interesting thinkers of interspecies communication. Marine mammal communication cannot be broached without the powerful concept of whale song, for example. Meanwhile, singing and music are frequently invoked in discussions of animals and aliens, by everyone from Charles Darwin to Carl Sagan. Love and desire are also crucial. Although supplementary to what is ostensibly my main concern (communication), love and desire supplement communication in the way identified above—they are *necessary* supplements, offshoots that are required for the life of the stem from which they grow.

Now for the layout of the book. I recapitulate and elaborate much of what I've said here in the introductory chapter that follows ("Interspecies Prattle"), although in a more granular and less abstract fashion.

The story really begins in chapter 1, which focuses on the individual who is most closely associated with interspecies communication: American scientist John Lilly (1915–2001).[8] The chapter shows that Lilly's attempt to dialogue with dolphins was tangled up with whale song research, and that music, spirituality, and love shaped much of his thinking.

Chapter 2 turns to love directly. It looks at scientists (Lilly, Irene Pepperberg) and feminist theorists (Donna Haraway, Alexis Pauline Gumbs) among others in an attempt to understand how desire, sex, and intimacy shape interspecies communication, whether explicitly or implicitly. Throughout the chapter, I keep an ear attuned to the animals who are loved, often in a cruel embrace, by humans. I emphasize the strange juncture between meaningful interaction and flight of fancy, the moment when "Sit!" becomes "I knew you'd sit because you love me."

Chapter 3 shifts to Lilly's other great interest (one might say his other true love): extraterrestrials. Using Lilly and American astronomer Frank Drake as points of departure, the chapter is an inquiry into the philosophical and psychological consequences of human exceptionalism. The idea that we are alone in the universe provokes tremendous existential angst. And existential angst is a paradoxical thing, when you think about it. Like cosmic loneliness, it is a dilemma at once very abstract and deeply personal. Chapter 3 looks at the anxiety, disorientation, and longing that is produced by a "condition" of human incomparableness.

The final chapter turns to a diagnosis of humanism in twentieth-century Black critical theory and art. In contrast to those who focus on the non-human side of interspecies communication (the animal or the alien), the figures discussed in chapter 4 offer a devastating critique of the human as a conceptual category. The chapter is a deep dive into the work of Sun Ra and Kapwani Kiwanga, artists who compel us to recalibrate our interspecies coordinates.

The book concludes with a brief epilogue that reflects on the lessons learned from four substantial chapters. I've also included a technical appendix on biosemiotics. It is my hope that readers without technical knowledge of animal behavior, communication theory, and music theory can manage this book. For those craving a more technical explication, I chunked details into this separate "chapterkin" (as one early reader termed it) so as not to overburden the rest of the book.

It's difficult to say much more so early in the book and without the necessary background, but a few additional points are in order. First, although the title and the framing of the book is "interspecies communication," I am interested primarily (almost exclusively) in human/non-human relations, and not those between other beings. The title may therefore seem misleading, but it is my contention that we cannot examine other relations (between birds and gibbons, for example) before better understanding our human relation to others. This is not because any attempt to comprehend non-human relations amounts to fantastical projection, but because the sciences as well as the arts often get tripped up on concepts that are generated from a human perspective. These concepts include language, meaning, and the notion of communication itself. I hope that this book will be useful for future work on interspecies communication in the more general sense. (In the sciences, the subject is sometimes termed "interspecific communication.")

Second, there is a politics to this book—a progressive, anti-sexist, and anti-racist politics. Is this book "overdetermined" by its politics? Perhaps. If so, I am OK with that. But politics is not moralism. Is it virtue signaling to critique a racist and misogynist author? Hardly. Where I am critical of someone, that criticism is only meaningful to the extent that it tells us something about their thinking and the context of their work. For example, a detailed account of John Lilly helps us understand the paradigm he was operating in; it helps us, in a way, produce an archaeology of marine mammal science through a single person. Part of that account will be critical, yes, but not in the sense of standing in moral judgment. Similarly, when I emphasize the colonial dimension of much thinking in the extraterrestrial life debate—from Fontenelle in the seventeenth century to Robin Hanson in the twenty-first—I do so to show how conservative political commitments make certain thoughts unthinkable.[9]

The book also discusses figures I admire and from whom I've learned. These people are "good" examples (or "good objects" to use my friend Tom Levin's turn of phrase). But I do not claim that scientists are bad and musicians are good *stricto sensu*, or that Afrofuturist artists are necessarily more progressive than animal behavior researchers. Perhaps "on average" they are (whatever that would mean—the idea is absurd). But that is not my argument in this book. Again, my approach is critical rather than moralistic.

Finally, I did not write this book to show that when we humans try to speak about animals or aliens we are really talking about ourselves. (I don't believe that to be true, at least not necessarily.) Nor did I write the book to demonstrate that we need not feel loneliness or existential dread because the world around us is filled with loquacious and melodious companions, from the bubbling brook to the magpie at my window. (I find that sentiment problematic, even galling.) If there's a message in this book, it's that we can and should try to communicate with the others around us, because those others are and are not like us, and that, for better or worse, we share the planet with them. The other message would be that only a slow and vigilant critical procedure combined with the best that animal behavioral science has to offer will get us anywhere at all.

If there's an ethos that animates this book, it's one that is idiosyncratic

to me. I've never been impressed with pronouncements about the wonders of the world and of nature. The world can be wonderful, yes, but it can also be bad, violent, or just plain boring. What has always amazed me is the wonder of thought. I hope that wonder will be clear on every page of the book you are about to read.

INTRODUCTION

Interspecies Prattle

In Paris in 1798, an orchestra and chorus from the Conservatoire de Musique readied themselves for a performance of works by Rousseau and Haydn, among other composers. But if the performance and repertoire were standard for the time, the audience was not. Where an eclectic audience of Parisian music lovers would typically be seated, now stood two elephants newly arrived from colonial Ceylon. The audience of humans at this particular concert, for there was indeed such an audience as well, had come not for the Rousseau or the Haydn, but rather for a much rarer spectacle of witnessing elephants respond to music. And what a spectacle it was. According to contemporary accounts, the elephants swayed their trunks rhythmically to the music and then, in a kind of bizarre denouement, began lustily caressing each other to the strident tones of a French revolutionary song. Go figure.

I am not the first to recount this tale. On the contrary, it is something of a staple in music histories of the period. James Johnson was perhaps the first to trot the story out in his now canonical book *Listening in Paris*, and the elephants have made multiple cameo appearances in the years since. And yet, as Nicholas Mathew and Maryann Smart observe, the story of the musical elephants only ever functions as a historical curiosity pressed into the service of some other purpose. For Johnson, the episode illustrates the capacity of music in post-Revolutionary France to civilize and regulate behavior in humans and animals alike. For another music historian, the story pithily captures the conjuncture of music and sex in late eighteenth-century Europe. In yet another recounting, the author John Deathridge presents a panoply of striking episodes (ranging from the elephants to a fabled musical spider that allegedly inspired Beethoven) to make an altogether different argument about musical community at the time. In short, the story of the elephants only ever appears as a brief, rousing episode, what Mathew and Smart describe as "a historical quirk."

As a quirk, its value lies primarily in the "thrill of encountering the unexpected thing."[1]

What would it mean to examine the story of the elephants not as a quirk, not as some titillating anecdote in a constellation of anecdotes, but as a central episode in music history? My project is motivated by this question, as well as by something that it brings to the surface: music's frequent use as a tool to overcome (or at least attempt to overcome) the putative boundary between humans and other beings. In other words, before matters of stylistic development, instrumentation, and other typical musicological concerns, musical modernity is arguably a matter of helping establish humanity's coordinates within the cartography of the cosmos. What do humans share with the various non-human entities that make up our planet and that exist in the wider universe? What are the possibilities for trespassing the border between ourselves and the world that surrounds us?

We know today that elephants do not and cannot respond to particular musical meanings, especially none as nuanced as those assumed by the Parisian audience in 1798. Here, it is important to not treat language too casually. In his recent book *Music, Math, and the Mind,* David Sulzer writes, "Do elephants have any comprehension of music? We read that they do. Rickye and Henry Heffner at the University of Kansas used a simple food reward experiment to elicit an Indian elephant's ability to distinguish simple, two-note melodies. They found that elephants could distinguish microtonal pitch gradations smaller than the half-steps of a piano."[2] That elephants can distinguish relative pitch with such precision is fascinating, but what musician anywhere in the world would describe a two-note sequence as a "melody"? The leap from a narrow, laboratory-based observation to a grand claim about animals and music is a common one in the literature, and one this book will carefully guard against. (Later in this introduction I will discuss how we can do better.)

If elephants cannot "comprehend" music, they certainly possess rich forms of sonic communication that are worthy of careful study—far more worthy, in my view, than making them listen to two sequential pitches, or trying to get them to play our musical instruments, as Sulzer has. We know today that elephants communicate by vocally emitting low sounds (below the 20 Hz range of human hearing) and that these rumbles can be heard by other elephants well over a mile away. Some scientists hypothesize that the infrasonic sounds are transmitted through the ground and detected by the soft padding of the animal's feet. Elephants use these rumbles, as well as other sounds, to communicate fairly specific content (signaling the need to get going, attempting to quell anger, warning about approaching danger), much as many large-brained mammals do. Note well: this is a form of communication, but it is not language in the human sense. To say that this form of communication is "elephant language" is as meaningless as saying that the ability to detect the difference between two frequencies indicates an elephant's capacity for music. It is worse than meaningless: it needlessly anthropomorphizes animals who have their own stunning behaviors, and, in its sloppiness, it does a disservice both to serious scholarship and to elephants. A more rigorous account of music, language, and animal communication will be needed to say something meaningful.

But Sulzer, almost despite himself, does make one interesting observation. In the course of an unsatisfying discussion, in which he describes how he and the scientist Richard "Professor Elephant" Lair built an orchestra for elephants to bang away at in rural northern Thailand, Sulzer adds as an aside: "Richard told me that the elephant's mahouts [i.e., people who tend elephants] know that elephants like to listen to music; they often sing to or play an instrument for the elephants as they walk together through the jungle, and the elephants are calmed."[3] We hear nothing from the mahouts themselves, nor do we have any evidence that the claim is true. (Maybe the elephants are calmed simply because the riders themselves are calmed by music—who knows?) But the mahout's claim, at least as relayed to us by Richard Lair via David Sulzer, is not far-fetched either. If indeed elephants are calmed by the mahout's song, but if it is nonetheless meaningless to talk of elephants comprehending the nuances of human music, then what exactly is going on? This is exactly the kind of question that this book addresses. And it does so without claiming (and without having to claim) that other animals appreciate music, or that they have language just as we do. What an elephant experiences when a mahout sings is no easy thing to explain. But it is surely worth some serious effort.

Let us return now to those elephants in Paris, for to leave them prematurely would be to treat them merely as curiosities. Their story tells us much else about the relationship between humans and animals, not least of which is the historically fraught definition of humanity itself. Even a cursory look at the history of how humans have treated animals reveals that some humans—and particularly Europeans—have always treated other groups of humans as quasi-human, or as not-fully-human. The Paris elephant scenario, after all, took place at a time when the French were conducting similar kinds of experiments on their colonial subjects in the very places where the elephants were captured. Europeans in the late eighteenth century wanted to know whether "natives" were capable of civilization and of reason, just as they had a few centuries earlier asked whether New World "savages" had souls and were capable of salvation. Elephants, of course, were more frequently killed for their tusks than treated to the music of Haydn. And Europeans of the time were far more likely to capture and sell a different commodity: an African slave.

The modern history of human/animal relations is mediated by, and saturated in, histories of colonialism and slavery. Questions about who and what is capable of language, reason, or the capacity for beauty have long encompassed animals, humans, and those humans deemed not-quite human.

Eighteenth-century European thought bequeathed to us many enduring ideas, one of which is the notion that climate—what would later be known as the "environment"—shapes behavior.[4] This idea was and is responsible for the racist notion that the frigid weather of Europe leads to increased rationality and decreased sentimentality, while the those living in hot or tropical climes lack rationality and are (over-) emotional. (Although few today would say this explicitly, the notion certainly persists.) But theories about climate influence were not limited to the domain of human culture. Bolstered by Enlightenment attempts to map out the entire world, and armed with a new emphasis on empiricism in science, European thought since the eighteenth century has attempted to comprehend all forms of life as products of the environment.[5]

Even before the eighteenth century, European writers had obsessively compared living forms ranging from humans and elephants to spiders and fish. Many such accounts appeal to the "worlds" in which those creatures live, with comparisons between humans and animals falling neatly beside comparisons between Europeans and non-Europeans. Consider, briefly, the work of Bernard le Bovier de Fontenelle (1657–1757), a French writer renowned for making scientific concepts accessible to a wider audience. In his *Conversations on the Plurality of Worlds*, he writes innocuously that "water is the atmosphere of fishes; they never pass into that of birds, nor the birds into theirs: they are not prevented by the distance, but the existence of both depends on their proper element."[6] If this is innocent enough, look now how effortlessly Fontenelle transposes those concepts both to people and outward across the cosmos: "What must the inhabitants of Mercury be? We are above twice the distance from the sun that they are. They must be almost mad with vivacity, like most of the negroes, they are without memory; never reflecting; acting by starts and at random: in short Mercury is the bedlam of the universe."[7]

Elsewhere in the same book, he notes that, while there are probably humanlike inhabitants (*habitants*) living on the moon, "I don't believe there are men [*hommes*] in the moon. We see how much all nature has changed even when we have traveled from here to China; different faces; different figures; different manners; and almost a different sort of understanding: from here to the moon the alteration must be considerably greater. When adventurers explore unknown countries, the inhabitants they find are scarcely human; they are animals in the shape of men."[8] Fontenelle is typical of a vein in European writing where a comparison between human and fish shifts frictionlessly in all directions, gathering up human difference and extraterritoriality in one fell swoop.

Such capacious comparison continued into the second half of the twentieth century (as it has until the present day), when scientists in many parts of the world developed something of an obsession with intelligent life in the oceans and in space. In both the United States and the Soviet Union, Cold War–era scientists assumed a parallelism between outer space and the "inner space" of oceans, which were conceptualized as new frontiers of knowledge, as realms that had previously been inaccessible to empirical inquiry but had finally become open to scrutiny with the aid of new technologies.[9] Jane Goodall's research on gorilla communication in Tanzania, John Lilly's work on human/dolphin communication in the Caribbean, and the first symposium on SETI (Search for Extraterrestrial Intelligence) all took place during the first few years of the 1960s. All of this work, moreover, made constant appeals to human cultural difference—dolphins, for example, were frequently equated with "primitive" people, since they were thought to possess some form of consciousness yet remain uncorrupted by technology and tools. Animals were also frequently compared with aliens, who were, in turn, compared with non-European peoples. The American scientist John Lilly (a figure I return to repeatedly) was motivated to communicate with dolphins because it would prepare us for alien contact. Once communication with dolphins was established, he wrote, they would for a long time "be in the position of the Negro races in Africa who are attempting to become Westernized," and who must "prove their usefulness in those things which the human races in general attempt to achieve."[10]

From the preceding discussions, one can derive a few general ideas. It is impossible to speak about the animal without a set of assumptions about the human. It is impossible to attempt to communicate with an animal without an implicit understanding of who we are in the course of that encounter. And it is probably impossible to understand our relationship to animals without some assumed notions regarding the limits of animal life: the alien, or at least the possibility of its existence, is always just beyond the horizon, shaping how we conceive who we are and who we are not. On this rocky terrain, language, music, and other forms of expressivity cross and double-cross lines that separate and ones that that connect. Those lines belie an impulse to trespass the border between the human and non-human, to discover humanlike non-humans either right under our own noses or at the far reaches of the universe.

✷

On the day that my mother died, my cat had a heart attack. I was sitting at the dining room table when my brother phoned to tell me that our mother had died. At that moment, my subjectivity seemed to crack, which resulted in my body letting out a wail. Helen, my partner, enveloped me in her arms. A few feet away from where I sat, Tigger went into cardiac arrest and toppled over—a soft, sideways thud. Everywhere, our lives and deaths are tangled with diverse kin. It was only a week earlier, after all, that Mom had phoned and, after a brief hello, asked to see Tigger on WhatsApp video. The vectors connecting us are multiple and tangled, moving in different (and sometimes opposite) directions, mediated by technology and history.

We humans form deep, needy, playful, generous, impatient, and violent relationships with the animals in our lives. We respond to them (when we know how to listen), and they—sometimes—respond to us, to our faces, our voices, our instructions. When we are lucky, they respond to our loneliness, and sometimes our devastation is too much for them to bear. But the range of human/animal communication is wide, from the spectacular to the quotidian and the instrumental.

Dogs synchronize their movements (including the pace of their gait and the direction of their gaze) with human companions. They also "recall their owner's face upon hearing the owner's voice." It seems that human yawns are "contagious" to dogs.

From the craggy alpine slopes to the plateaus of Central Asia, shepherds corral flocks using specific, sometimes ancient melodies. (Some of these melodies have been thematized, for example by Rossini and Berlioz.)

Parrots are wonderfully loquacious creatures and have proven treasured companions for seafarers out on silent, endless waters. People who spend time with these birds know that they are capable of far more than imitation. Some parrots are able to identify numbers, shapes, and colors when asked by doting owners; these same birds may ask to go home when quizzed too many times. (These results remain controversial, however.)

Snake charmers neither charm nor mesmerize snakes. Instead, they compel snakes into a state of ongoing terror—the snake responds to the movements of the "charmer's" wind instrument as if the slender piece of wood is a predator that might attack at any moment.

In the northern most part of Mozambique, people know that a rather ordinary looking bird with a peach-colored beak will lead them to honeycombs when solicited. Human honey collectors call these birds using a specific sound (something like "brr-hm!"), and they seem to understand that the birds respond especially well to that particular vocalization.

Birds, of course, are excellent students. They learn intricately structured "songs" from their parents (a few sparrows have been taught new songs by humans, as well). For centuries, humans have been mesmerized by the vocal abilities of our soaring companions, who have been the subject of musical compositions, art, and theoretical treatises. In Surinam, men enter their songbirds into competitions held in parks: the bird that sings longest wins, and their owners spend untold hours trying to figure out how to nurture vocal endurance.

From Australia to Mexico, horse riders and horses attune themselves to each other's micro-movements since, as one writer famously put it, "Every muscle twitch of the rider will be like a loud symphony to the horse."[11]

All over the world, people communicate with animals—at least if we understand communication in its most basic, etymological sense as partaking of something in common.[12] The above are all examples of vernacular knowledge, instances of "know-how" rather than "know-that": shepherds know how to call flocks, honey hunters work cooperatively with honeyguide birds, and so on, but these relationships are seldom subjected to explicit theorization, and even less often to controlled empirical study.[13]

What happens when we consider the scientific field of animal cognition, where knowing-that always trumps knowing-how, and where a revolution has been taking place since the second half of the twentieth century? Researchers have discovered that animals are far more cognitively complex than dominant scholarly paradigms have assumed, and today it is not uncommon for researchers to insist that many species of non-human animals possess a rich array of cognitive capacities, or even consciousness. Although baboons can vocalize only a few sounds (three or four kinds), they live in hierarchical social structures, so even an exchange of two simple sounds requires complex interpretation since the matter of *who* is responding to *whom* with *which* sound matters. Cuttlefish seem to be able to delay gratification when something better is on offer down the road. Humpback whales transmit complex "songs" across the ocean, and they continually modify those songs in ways that we are only beginning to understand.[14] The startling world of animal behavior is coming increasingly into focus, even if much of it remains downright mysterious.

What is the connection between the vernacular know-how of dog lovers or horse riders and the scientific jargon of academic journals? In many instances, discoveries about animal behavior made by scientists at the Princeton Neuroscience Institute, the Max Planck Institute for Animal Behavior, and Cape Town's African Institute of Ornithology are only affirming what people who interact with animals in their everyday lives already know. This, indeed, is a point often emphasized by animal researchers. Here is Jane Goodall speaking at a recent, highly publicized event at MIT:

> The subject of animal intelligence is so absolutely fascinating. And of course, as all of you probably remember, it was when I went to Cambridge University in the early '60s—I was told animals didn't have minds that could solve problems. And that we humans were completely unique. So, as I had a wonderful teacher when I was a child, I was able to stand up to those professors—which was pretty brave of me because I hadn't even been to college. And there I was doing a PhD. [Goodall reaches over to pick up a photo of herself as a young women seated next to a dog.] And this was my childhood teacher, Rusty. And, as all of you know, animals are intelligent.[15]

The sentiment behind this story (which Goodall has repeated countless times over the years) is a common one. A number of scientists have attributed their basic recognition of animal intelligence to a childhood relationship with a pet or another animal.[16] Animal intelligence, in other words, is less a recent *discovery* than the overturning of a particular intellectual tradition in the name of animal lovers everywhere, and in the name of animals themselves. While Goodall is able to make pronouncements about animal intelligence only after rigorous study, she nonetheless continues to pay tribute to Rusty, not as her primary research subject, but as her most important teacher.

In sum, we know that animals are intelligent, and we know that communication with them is possible. But intelligence and communication are two of the least precise and most loaded words in the English language. The difficulties start when we try to define intelligence and when we try to explain what is happening between animals and ourselves in the course of "communication." One of the central tenets of this book is that attempts to describe or explain human/animal communication often fall back on, lead to, or even generate other, more flexible (and often more elusive) concepts. Having spent the past several years immersed in this topic, I have noticed several concepts that frequently emerge in or from attempts to characterize our communicative endeavors with animals. One is music. Another is love. (I will talk about a couple of others in the course of this book.)

Recall the quote above about "every muscle twitch of the rider be[ing] like a loud symphony to the horse." Another writer insists on a "silent language" that exists between riders and their horses.[17] Because they cannot say what it is actually like to communicate with a horse, these writers resort to metaphors—of music (symphonies), on the one hand, and of (silent) language, on the other. Whatever happens between horse and rider is some strange form of communication sandwiched between music and language but not equivalent to either.

The turn toward such concepts typically arrives at the end of an attempt to explicate the mechanism of interspecies communication. In some cases, a concept like music acts as a kind of quasi-explanatory aid, naming what has happened in an intuitive and only partially satisfying manner when all other attempts have failed. But concepts like music (or love) are not *merely* stand-ins for proper explanations; they do not merely fill in a lack. Rather, we might say that any attempt at interspecies communication—whether it is "successful" or not—generates something else, something more. For years, researchers attempted to communicate with cetaceans (that is, dolphins and whales). They didn't succeed in the way that they had hoped, but they "discovered" in the process that whales sing. While scientists continue their attempts at simple interactions with captive marine mammals (dolphins are large-brained mammals, so it comes as no surprise that basic communication is possible), the experience of listening to recordings of whale song has proved far more impactful and profound. It is not uncommon, either, for researchers working with dolphins to declare their love for the animals, even when those same researchers have no lexicon whatsoever to explain their passion. So, too, with dogs and cats. Every dog lover knows that a vast world of communion exists beyond instructions to sit or fetch or rollover. (Pet owners may intuit some of what scientists are only recently proving about our canine companions—for example, that dogs differentiate between human words, or that they synchronize their gait to our own.) But despite struggling in vain to describe communication beyond basic instructions, many dog owners would admit their love and devotion to the canines in their lives. How can we understand this tremendously complex terrain? Let's look at a few approaches to the topic. Although the field is still underdeveloped, several strands of inquiry do exist.

One strand consists of narrow, highly controlled scientific studies. Studies of animal pitch perception are a case in point. "Humans Identify Negative (But Not Positive) Arousal in Silverfox Vocalizations" reads the title of another recent article.[18] Another (slightly earlier) study observes that shepherds use "short, rapidly repeated notes, with a tendency to rise in frequency" when attempting to stimulate border collies and that, because non-human primates and birds use similar "acoustic structures," employing those structure/s may increase the likelihood that canines will respond as desired by herdsmen.[19] This strand, in brief, focuses on small pieces of the puzzle—like much of the best scientific research. And because the field is still in its infancy, researchers in this area typically refrain from making strong claims. The authors of the silverfox study, for example, admit to not knowing why humans respond to negative but not positive arousal. The conclusion of the shepherd article is tentative, relying as it does on a flimsy comparison between the calls of herdsmen and birds. It goes without saying that this kind of patient, narrow work will continue unabated for a while yet.

The most eye-catching work is that where non-human animals are taught "artificial" languages. The best-known examples are apes who are taught sign language, with Koko the gorilla and Washoe the chimpanzee reaching near-celebrity status.[20] Another famous example is the African grey parrot Alex, whose ability to detect shapes and colors (for example) seemed to overturn the long-held belief that parrots are merely imitative creatures. These studies—along with work on dolphins, cephalopods, and other animals—have been crucial in cracking claims that animals are mere automatons. (Once again, this belief is more prevalent among scientists and academic philosophers than it is among people who share everyday experiences with animals.) But these cases have also generated a number of other concepts that have spilled or fallen out of scientific inquiry.

At the MIT event mentioned above (the one where Goodall talks about her dog-teacher, Rusty), opening remarks were provided by none other than British musician Peter Gabriel. Gabriel began by showing a video of a performance with a chimp: the chimp hits a bunch of white notes on a piano that sometimes seem to make musical sense with Gabriel's ethereal C major synth backing. (To anyone who knows anything about music, it's obvious that there's no way this effect could *not* happen.) Following the video, Gabriel makes some remarks that illustrate how flexible concepts are often used to bolster claims for, or to serve as a quasi-explanation for, human/animal communication. He begins by speaking about the chimp-pianist: "It blew me away. And then watching the way that they were using symbols for communicating and then, with Penny [Patterson] and Koko [the gorilla], seeing how the sign language was used. And the sort of poetry came out [of] the combination of words. It led me to think that you'd have to be deaf, blind, and very dumb not to notice language taking . . . being used. So I got very excited about that."[21]

While (and perhaps because?) the capacity for combinatorial complexity among gorillas remains a highly controversial issue, Gabriel invokes the notion of poetry, which obviously functions as a metaphor for something he is not able to adequately name. And then, after his strangely ableist comments, he launches into a rant that touches on the Kabbalah, consciousness-raising, and his interactions with wise "indigenous people." Discourse around human/animal communication is frequently a mélange of narrow, positivist claims (recall those silverfoxes) and the conjuring of other concepts.[22]

A related strain appears in the social scientific model pioneered by Donna Haraway, Bruno Latour, and others. This work has an ambiguous relationship to science: while it selectively deploys claims made by lab-based researchers and ostensibly supports mainstream scientific practice, it is also *about* science in a manner that some interpret as critical or even skeptical. Perhaps the strongest and most general argument by authors working in this area has been that the human is hopelessly entangled with non-human Others. Indeed, we ourselves are partially made up of and sustained by microbes. Even more intriguingly (and more complexly), endogenous retroviruses constitute as much as 5% of the human genome, which suggests a significant presence of non-living or only quasi-living retroviruses in our deep evolutionary development. In a recent article on retroviruses and stem cells, Jennifer Johung writes provocatively, "Becoming attuned to the processes that generate and re-differentiate life also leads to the recognition that non-life is dependently attached to the very hallmark of renewable life, in all of its complexities and varieties."[23] The message, in short, is that any attempt at separating humans from non-humans only results in further human/non-human hybrids. We cannot live without the splendid universe of animals, plants, bacteria, and other biological entities that make up our bodies and world—we can only ever live in and with relation, in and with connection.

Haraway in particular has focused on the ethical and political consequences of this insight, and has long insisted on our responsibility to and for the creatures who co-constitute our lives. Similar to people like Peter Gabriel, yet in a different way, Haraway mixes discourses—because she is so committed to hybrids and entanglement, this could hardly be otherwise. Communication, then, never takes center stage in her work but vies for attention with other concepts in a way that should by now sound familiar. Consider, for example, Haraway's discussion of *My Dog Tulip*, a book by British author J. R. Ackerley. In keeping with her style, Haraway begins off-center, with a nod in an entirely different direction:

> The Dutch environmental feminist Barbara Noske, who also called our attention to the scandal of the meat-producing "animal-industrial complex," suggested thinking about animals as "other worlds" in the science fictional sense. In his unwavering dedication to his dog's significant otherness, Ackerley would have understood. Tulip mattered, and that changed them both. He also mattered to her, in ways that could only be read with the *tripping proper to any semiotic practice, linguistic or not.* The misrecognitions were as important as the fleeting moments of getting things right. Ackerley's story was full of the fleshy, meaning-making details of worldly, face-to-face love. Receiving unconditional love from another is a rarely excusable neurotic fantasy; striving to fulfill the messy conditions of being in love is quite another matter. The permanent search for knowledge of the intimate other, and the inevitable comic and tragic mistakes in that quest, commands my respect, whether the other is animal or human, or indeed, inanimate. Ackerley's relationship with Tulip earned the name love.[24]

In this bracing passage, Haraway moves from environmentalism to a science-fiction-like notion of other worlds, before making a brief stop at the pun "significant otherness" on the way to a tentative destination: semiotic (but not linguistic!) practice. (Having read the book many times over the years, I still trip over the phrase "the tripping proper to any semiotic practice"—what does Haraway mean here?) But semiosis is a brief stopover, at best, for Haraway takes off again, touching on misrecognition before arriving at what seems to be the real subject of this passage: worldly, face-to-face love. This grounded, unabstract love commands respect, says Haraway, before adding that this kind of love (as well as the respect it commands or perhaps even demands) is possible even with inanimate things (as a reader, the last term surprises and intrigues me).

For better and worse, Haraway's writing does not say much about communication itself, because it does not and cannot dwell on that (or any) topic for long enough to produce a sustained analysis. This, of course, is deliberate, and it is somewhat effective (which is why I said both better *and* worse). But, and here I take Haraway as a paradigmatic figure of the field, the social science literature on human/animal communication typically leaves me wanting, in large part because communication itself remains somewhat elusive and undertheorized.

I take two lessons from the brief reading of Haraway. First, a more robust analysis of human/animal communication—one that does *not* immediately jump to love and respect and other issues—could be fruitful. Second, even though Haraway's writing is pretty extreme in this regard, it is true that *most* writing on human/animal communication seems to veer off, at some point, into alien territory. As I have suggested, the topic of human/animal communication provokes or incites a restlessness, a pull toward a theme like music, or poetry, or love, or ethical commitment.

In the preface, I pointed out a similar tendency in the work of John Durham Peters, who values notions like cooperation and stewardship over semantic or dialogic exchange.[25] All of this is fine—even admirable, from an ethical perspective. But communication as such has a tendency to get lost along the way. This book aims for a better balance.

A separate body of literature, less academic in nature, deliberately anthropomorphizes non-human animals, partly because of an irresponsible use of language and partly for shock value. "My starting point is that animals have language," writes Eva Meijer in *Animal Languages*.[26] Peter Wohlleben's *The Hidden Life of Trees: What They Feel, How They Communicate—Discoveries from a Secret World* fails to mind the same gap—fails, that is, to recognize the difference between arboreal homeostatic processes and something more complex (and meaning-laden) like "feeling."[27] Both books are easy to read, brim with intriguing anecdotes, and fulfill the need of many contemporary readers to reconnect with nature. In a similar way to work described above, both books shift seamlessly between human and non-human, between emotion, love, and care. But where Haraway keeps the semiotic aspect of her work murky and difficult to pin down, and where Peters shifts *from* semantics *to* ethics, Meijer and Wohlleben build their ethics on a capacity for language and emotion that humans (in their view) share with animals and trees. I find work in this category unscientific as well as politically dubious.

Other approaches exist. Probably the most fruitful among them is a particular take on "biosemiotic theory." Biosemiosis—at least according to the authors I look to for inspiration, most notably Gary Tomlinson and Eduardo Kohn—refers to the use and interpretation of signs by many animals, including but not limited to humans. It welds together Darwinian evolutionary models with a theory of signs developed by the American philosopher Charles Sanders Peirce (1839–1914). Current biosemiotic theory has several things going for it. It is grounded in empirical research, but it is also deeply theorized and in contact with movements in critical theory. It is, on this score, both scientific and humanist, and, in my view, takes the best of both traditions. At its best moments, biosemiotic theory is attuned to and concerned with important political questions as well—something that cannot be said of all scientific fields. As a bonus, biosemiotic theory provides important insights not only about language and communication, but about music as well.

This book includes an appendix that covers the thornier aspects of biosemiotics. There, the reader will find a rigorous, if brief, explanation of the theory (or at least those elements of the theory I find most valuable). All I wish to emphasize here is that biosemiotics can, to a surprising degree, account scientifically for how different meaning-making systems function. It does this by starting with the building blocks of semiotic systems, rather than beginning with systems as fully fledged and impenetrable. Biosemioticians are keenly attuned to the overlap and continuities between semiotic structures, while also shedding light on how they differ. Biosemiotics (at its best) refrains from facile, negative definitions that are unfortunately all too common in the literature: it offers positive definitions rather than simply declaring, for example, that music is *more* emotional than language, or that animals communicate *less* abstractly than humans.

The work of Tomlinson in particular helps us understand what we lose by applying terms like *music* and *language* willy-nilly to scenarios where they are not appropriate. Following Tomlinson, in this book I guard against casual pronouncements such as, "Birds have language since they communicate," or "The cries of a gibbon have musical cadences and are, therefore, a kind of singing." Although the focus of this book is more critical-historical than scientific, it seems to me that any discussion of interspecies communication requires a rigorous deployment of key terms. Not all communication is linguistic (which is why we should be leery of applying the word *language* to birds), and we can speak meaningfully about gibbon cries (including an aspect like pitch-contours) without likening them to either music or singing. This is the kind of careful thinking in which authors like Tomlinson and Kohn are engaged.

If Tomlinson and Kohn help us see the distinctions between various forms communication, they also help us identify a region of semiosis shared between humans and other animals. (This region of semiosis is principally made up of indexes—signs concerned with pointing or causality. Many non-human animals use these for a variety of purposes.) Kohn, who deploys biosemiotic theory in his ethnographic work on Amazonian communities, refers to communication in this overlapping meaning-space as "transspecies pidgins."[28] Because it is based on empirical research with animals, and because it provides a lexicon for different kinds of communication and expressivity, biosemiotics provides an antidote to those (typically dyed-in-the-wool humanists) who insist that we must remain agnostic about anything other than symbolic communication.

But despite its potency, a strictly biosemiotic account of human/animal communication has one serious limitation: even if we are able to rigorously establish the continuities and boundaries between human and animal semiotic systems, humanity is not and has never been defined in purely semiotic terms. The human, in addition to being a species, is also complexly layered sociopolitical being, where value (i.e., whose life matters, or who counts) is a key consideration.[29] Philosophers have variously characterized humans as beings with a soul, with language, with rationality, with moral virtue, with free will, or with the capacity to change and develop—to list just a few monstrously thorny examples. None of these terms are definable in value-free language, which is why, historically, certain human population groups have been excluded from the category of the human (or what we might call the Human). Aristotle famously made a distinction between voice (*phonē*), which is shared by humans and animals, and rational speech (*logos*), which is a distinctly human attribute. But he also qualified which humans have the capacity for rationality, arguing that slaves do not.[30] In other words, no matter how powerful it is (and it is powerful), biosemiotics does not and cannot provide an all-encompassing definition of humanity. Nor can it offer a comprehensive account of humanity's relationship with the non-human world.

Biosemiotics does, however, point to a sphere of semiotic activity that various kinds of creatures have access to. Even if scarcely any species other than the human possesses the capacity for symbolic meaning (or, for that matter, music), a great many creatures on our planet participate in a world of indexical (and perhaps iconic) semiosis, and it is there that interspecies communication flourishes.

This book, therefore, moves continually between a more scientifically driven approach informed by biosemiotics, and a political orientation informed by history and critical theory.

✷

When I was writing the book you are now reading and was pitching it to various people, I often said something like, "Well, the book is about interspecies communication, but it's not like I am going to write about Dr. Dolittle!" But reading the books (for I had not read them yet when I made those pronouncements) changed all that. Simply skimming the contents page of the first volume, I quickly realized that Lofting's book was a kind of distorted mirror of my own.

Dr. Dolittle presents a dreamworld where a British man understands, communicates with, and is capable of saving all the living creatures of Earth. Written and published in serialized form by Hugh Lofting between 1920 and 1954, the Dr. Dolittle books have long been criticized for their rampant racism, the most famous example being the story of the African Prince Bumpo, who pleads with Dr. Dolittle to make him White. Accompanied by a drawing of Bumpo (done by Lofting's own hand) that would fit on the vilest Blackface minstrelsy poster, the story is so offensive that even diehard fans have supported its redaction in recent editions. But editing out that particular episode, welcome as it is, cannot rescue a book so thoroughly saturated in racism. From the African Bumpo—whose head was so hard, we are told, that when a bottle hit it, his head was fine but the "bottle smashed into a thousand pieces" (a passage *not* redacted in later editions)—to the many ridiculously stereotyped indigenous people and tribes (examples include the noble Long Arrow, in touch with plants and Nature, and the eternal war between the idiotically named Popsipetals and Bag-jagderags) and the fact that a British man of science is summoned literally all over the world to save all manner of creatures, from indigenous people and Africans to birds and crustaceans, Lofting's Dr. Dolittle books are beyond redemption.

Ostensibly, at least, the books are about communication. Dr. Dolittle communicates with everyone and everything. He can speak all human languages (including "dead" languages), as well as the languages of many animals. His world is a colonial one, shaped by the vast riches of the British empire in the first part of the twentieth century. He travels the world with a genuinely motley crew made up of a young boy (the narrator, Tommy Stubbins), an African, a chimpanzee, and an African grey parrot. The chimpanzee is named Chee-Chee—the same name, oddly, of

John Lilly's first talking dolphin. (I discuss Lilly's Chee-Chee in the next chapter.) Dr. Dolittle learns his first animal language from his African grey parrot, which the good doctor could only acquire, of course, because of colonial connections to Africa (frequently referred to as a "country" in the books). Curiously, Dr. Dolittle names the African parrot Polynesia, which implies, it seems to me, the broad scope of the globe at his disposal. (Side note: the fact that Dr. Dolittle learns an animal language first from an African grey parrot was not lost on the famous American scientist Irene Pepperberg, who ends her 2008 best-selling book with a romantic rumination on Africa. I discuss Pepperberg and her parrot in chapter 2.)

But there is more. The first Dr. Dolittle novel concludes with the doctor going *underwater* (by riding in the transparent shell of a massive snail) in order to learn the languages of deep-sea creatures. This episode relates to two ideas, prevalent in the history of human/animal communication, that I discuss at length in this book: (1) that attempting to communicate with marine animals is best done if we can find a way to spend time underwater ourselves; and (2) that we struggle to communicate with aquatic creatures because the sounds we emit travel differently through water than through air. Marine animals and the vagaries of aquatic communication are already there, in this famous children's book from the early twentieth century.

A later book in the series has Dr. Dolittle flying upon a massive moth beyond Earth's atmosphere—all the way to the moon. Even in Lofting's children's book, I learned, aliens make a substantial appearance. The noble doctor is called to the moon by its poor inhabitants, whose health depends solely on his generosity—without him, the creatures of the moon will suffer miserably and without respite. "Caterpillars as long as a village street with gout in a dozen feet; immense beetles suffering from an affliction of the eyes; grasshoppers as tall as a three-story house with crude bandages on their gawky joints; enormous birds with a wing held painfully in an odd position": Dr. John Dolittle heals them all.

And here is the most uncanny part: Dolittle's first attempts to communicate with non-human creatures on the moon take the form of music.

> There was no doubt about it at all: the trees were making the sounds and they were doing it *deliberately*. In the way that an Aeolian harp works when set in the wind at the right angle, the trees moved their branches to meet the wind so that certain notes would be given out. The evening the Doctor made his discovery of what he called the *Singing Trees*, he told me to mark down in the diary of the expedition as a Red Letter Date.

> I shall never forget it. . . . "Why, Stubbins," said he, "do you realize what this means?—It's terrific. If these trees can sing, a choir understands one another and all that, *they must have a language*. They can talk!"[31]

And so it was that Dr. Dolittle learned to speak with the trees, and later, with the animals of the moon. Dolittle's logic here is wonky at best, relying as it does on a quasi-non-sequitur. Singing trees must have language, because a choir (that is, a group of singers) understands "one another and all that": it's as though Dr. Dolittle got confused and gave up, ending the sentence with "all that." Alas, one finds such imprecision even in renowned scientific works. For many reasons, Dr. Dolittle's approach will not work for us, even though we will meet many Dolittle-like characters in this book's pages. Will we have the humility to acknowledge how history shapes the way we listen to animals? Will we have the courage to overcome narcissism? Do we have the patience to forego sentimentality and use language carefully?

Any interspecies inquiry must begin with these questions.

I

Animal

✷ 1 ✷
Lilly's Wager

According to experts on the topic, it is only a slight exaggeration to say that music saved the whales. Indeed, several species of whale were well on their way to extinction before Roger Payne and Scott McVay discovered whale song, or rather, before Payne and McVay taught the public to hear whale phonation *as* song. Despite the protestations of long-time marine mammal researchers, when it came time to make a case against the whaling industry, the gentle bellowing of a single male humpback some two thousand meters below the ocean surface proved more effective than careful argumentation.[1]

In contemporary biosemiotics terms, whale song is a set of patterned gesture-calls: humpbacks and other whale species vocalize "units" of sound in repeated and combinatorial ways over long durations. Recordings of these elaborate vocal expressions served as the source for a movement in the 1970s that reconceptualized whale sound as singing.

Payne produced the recording *Songs of the Humpback Whale* in 1970 and, with McVay, published the groundbreaking article "Songs of Humpback Whales" the following year.[2] When NASA opted to include an excerpt of whale song on the famous *Voyager* album that was sent into space, the president of the National Geographic Society, Gilbert Grosvenor, declared that "the whale has become a way of thinking about our planet and its creatures."[3] McVay made the analogy even more explicit: "To leave the oceans barren of whales is as unthinkable as taking all music away."[4]

Several scientists and historians have observed that song was the single important feature in the battle against the whaling industry, which was largely won when the Marine Mammal Protection Act passed in 1972, just a year after the publication of Payne's and McVay's first article.[5] In that 1971 article, Payne and McVay made the dramatic claim that whale phonation is composed of clear, repeating structures—what the scientists insisted were *songs*.[6] While the "haunting mewls and honks" of humpbacks had been known to scientists since the 1940s, these had previously been thought of as mere noise.[7] But now, in addition to the article in *Science*, Payne's 1970 album *Songs of the Humpback Whale* brought the music of these newly majestic creatures into people's living rooms.[8] In a few short years, the sounds of whales were transformed in people's minds and ears from mere animal cries into ethereal or even transcendent music.

This emphasis on whale song notwithstanding, the conditions of possibility for such a musically inflected political gesture ran deeper. Most whale historians agree on a deep transformation in the perception of whales roughly between 1960 and 1975: over the course of fifteen years, these creatures went from dangerous, even vindictive behemoths (desirable primarily for their blubber) to intelligent and peaceful creatures harboring the capacity to show us how to truly live.[9]

The transformation was the result of multiple industries, innovations, and political projects. It also often included marine mammals other than whales—especially dolphins. (In contemporary scientific terms, whales and dolphins are part of the order Cetacea.)[10] Following WWII, the United States invested millions of dollars in oceanic research, which included training dolphins to detect underwater mines and developing submarine sonar technology based on marine mammal echolocation.[11] This research took place alongside an explosion in American popular culture, including films like *Flipper*, *The Day of the Dolphin*, *Star Trek IV: The Voyage Home*, science-fiction literature like Leo Szilard's *The Voice of Dolphins*, and the development of a robust animal rights movement in the 1960s and '70s, including anti-whaling campaigns, such as those initiated by Greenpeace. These disparate factors coalesced to produce a coherent understanding of cetaceans. The period between 1960 and 1975, in other words, brought a dramatic epistemic shift.

A number of unpredictable conceptual linkages were also forged—for example, the symmetry between outer space and the "inner space" of oceans. Feminist theorists Metta Bryld and Nina Lykke write about how dolphins encapsulated the zeitgeist: "All of a sudden, these sea creatures jumped smilingly out of the blue as stand-ins for the citizens of the cosmic super-civilizations that possessed techno-godfatherly wisdom, and with whom it was believed the Space Age would bring us into contact."[12]

Although the dramatic shift in our perception of cetaceans is not reducible to any individual person or institution, historians have emphasized the role of John C. Lilly (1915–2001), an American researcher who is best remembered for his attempt to break through the human/dolphin communication barrier.[13] In the 1950s and '60s, Lilly was viewed as a renowned if highly idiosyncratic thinker. He won major research grants, and counted luminaries such as anthropologist Gregory Bateson and astronomer Carl Sagan among his closest friends. But consistent scientific failure (Lilly never managed to teach a dolphin to speak) and unorthodox practices (he administered LSD to his dolphins and himself) sent Lilly's research into a cul-de-sac. For the last decades of his life, Lilly was dismissed by mainstream science and lived out his years as some combination of activist, hippy, guru, and quack. Through it all, Lilly never gave up on his idea, his unshakable axiom, that communication with dolphins is possible. Dolphins would continue to attempt to speak to us, he insisted, even if we stopped trying to speak to them. His faith in dolphin intelligence and empathy hardly wavered for a second.

Although Lilly has long been dismissed in mainstream science, the ideas he promulgated in his 1961 book *Man and Dolphin* have since become mainstream ideas in Euro-American culture. But even in the domain of science Lilly's position is not so simple to assess. The historian D. Graham Burnett has argued that, despite the onward march of science, demystifying the Leviathan through observation and empirical research led unexpectedly to more mystification, to re-mystification.[14] Over the course of the twentieth century, Burnett observes, "the wild legends, the seafarers' yarns, the biblical tales, all gave way—fell away—as men of learning pursued the cetaceans with harpoons, microphones, nets, and neurological probes, leaving a pile of quaint old beliefs in shards (on the one hand) and a proper knowledge of what cetaceans were (on the other). And yet," Burnett continues,

> somehow, in the process, what emerged was a new creature of extraordinary symbolic power, whose looming significance swelled with each seemingly deflationary blow, whose new iconography was composed, rapidly, mosaic-like, out of the broken bits of old myths (so navy dolphins *were* being trained to help sailors, as in the ancient tale of Arion . . .), heightened with the bright shards borrowed from the workshops of scientific cetology (so the whale hugger-saboteurs releasing navy dolphins back into the wild could believe that the animals "saw" the fear and goodwill of their liberators by sonar-scanning their racing hearts.).[15]

As the brief allusion to Arion in the above passage implies, the history of human/cetacean relationships goes back a long way. The ancient Greek myth tells the story of musician and poet Arion, son of Cycleus, who was captured and nearly put to death by sailors. Arion escapes by playing his lyre and singing to dolphins in the nearby waters, who come to his aid and rescue him. The story of Arion, in other words, is a foundational interspecies communication myth. And while any such lofty thoughts about cetaceans seem to have disappeared during the brutal period of modern whaling (roughly between the seventeenth and early twentieth centuries), Burnett suggests that scientists in the 1960s tapped into remnants of those myths.

There is an argument to be made, in fact, that Lilly's imprint can still be felt not only in popular culture, but even (if only indirectly) in the most rigorous scientific discourse. Burnett concludes that, although Lilly's work has been maligned by the scientific establishment for decades, it was his mystical, quasi-scientific position that ultimately won out, that most contributed to anti-whaling activism, and that provided the framework for listeners to hear whale phonation as song. It was this framework, moreover, that inscribed whales within an ethical grammar of suffering and care.

This chapter focuses on John Lilly. I understand Lilly not as some quirky, minor historical figure who might, by virtue of his failures, help us understand hegemonic perceptions.[16] Lilly, rather, was a well-resourced researcher who, although he certainly took an unconventional turn, can nonetheless be viewed in historical hindsight as a spectacular success.

His success notwithstanding, Lilly was more than a bit naive. In some cases, his naïveté is vaguely charming, but in many more cases it makes the contemporary reader cringe. During the main period of his research, scientists were thinking deeply about the issues Lilly was tackling in an explicitly iconoclastic way.[17] Lilly was not working in a vacuum—even though he often acted that way. And therein lies the rub: Lilly's most misguided ideas became his most popular, entering even into the scientific domain. He was most successful when tapping into the deep wells of Euro-American ideology.

But significant historical lacunae remain. Burnett implies that Lilly laid the groundwork for the global whale song phenomenon, but the historian does little to explain the actual connection between Lilly's dolphins and Payne's whales, in part because Lilly himself seldom wrote about whales, singing, or music.[18] In this chapter, I return to Lilly's work in an attempt to tease out those connections. More generally, however, I consider the way that music has often mediated human perceptions of cetaceans. This is related to what Ana Mariá Ochoa Gautier calls the "ecologization" of music: that is, the scenario in which music becomes the "much-needed suture for the torn relations both between humans and between humans and the environment."[19] We see this use of music during the hippie movement of the 1970s, and today again in the deployment of music in New Age circles.

This chapter shows that Lilly's attempts at suturing the torn relations between humans and non-human animals were constantly frustrated by the actual media required to do that suturing. By "media," I mean anything between or linking two things—whether that be technological media (such as magnetic tape), communication media (like sound or vision), or environmental media (such as water or air). Much of our contemporary perception of marine mammals, especially as it has been bequeathed to us by Lilly, depends upon an impatience with and ultimately a disavowal of media. Music and water become important grist for the anti-media mill. Music and water become anti-media figures that purport or pretend to allow immediate connection—connection without mediation.

Lilly's Dolphin Hypothesis

Two episodes, separated by nearly a decade, served as the foundation for John Cunningham Lilly's dolphin hypothesis. In the first, a young Lilly, newly qualified as a medical researcher at the University of Pennsylvania, traveled to inspect the carcass of a beached whale in Maine. "I felt awed and diminished in size as I saw the huge stranded beast," he recalled, and he wondered how such a creature "lived, what it thought, if it talked to its companions."[20] He was especially impressed by the size of the animal's brain, which he observed "was very much larger than a human's," noting that its "fissures, gyri, and sulci (folds) were far more complicated."[21] After several years conducting gruesome, failed experiments on macaque monkeys, Lilly recalled his earlier encounter with the beached whale and turned his attention—in an almost obsessive fashion—to marine mammals. But his early research on dolphins, which he began in 1955, yielded no advances and a great deal more suffering to his test subjects, many of which died during his experiments.

The second major episode has been widely documented, including by Lilly himself. On a typically uneventful Sunday morning, as he poked around in a dolphin's brain, something surprising happened: the creature began making sounds he had never heard before.[22] Lilly recounts the build-up to this moment in gruesome detail (readers may want to note this before reading on). He began by applying the so-called "sleeve-guide technique" (which he had previously used only on monkeys) to dolphins: first, he had to restrain the dolphin (the animal's "body was supported in the water in a sling hung from two pipes. . . . The head was held down by a bent steel piece shaped to the back of the neck")[23]; after a local anesthetic (whose pricks caused the dolphin to jump repeatedly), he hammered the "sleeve guides into the skull" (71). For this, he required a carpenter's hammer after realizing that a smaller one would not do. With the sleeve guides in place, he then inserted electrodes through the epidermis and tissue, all the way into the brain. The whole gruesome experiment was an attempt to get inside the dolphin's head—literally and figuratively. It was as though Lilly was searching for some self, some soul even, within the recesses of the animal's body.

Having rigged up the animal, he was able to electrically stimulate specific regions of its brain, causing one of the dolphin's eyes to "turn in a definite direction and hold its directional gaze as long as the stimulus was on" (73). Even so, progress was slow since the brain of a dolphin was largely uncharted territory. Lilly felt, he said, as though he was "shooting in the dark" (73). After failing to elicit much beyond the eye tweak, he felt discouraged. "We thought maybe this brain was all silent," he recalls. "Our time was getting short and we were getting impatient" (74). One glimpses another side of Lilly's character (one we will have occasion to return to) in the sentence that follows the one just quoted: "Finally we were forced to leave the laboratory to attend a party that evening."

It must have been a groggy and hung over John Lilly who entered the lab that next morning. A John Lilly who, in his impatience, "cranked" the electrode deeper yet into the dolphin's brain and turned the current up to an unprecedented level (75). It was at this moment, we read in *Man and Dolphin*, that the dolphin emitted a "more exuberant vocalizing" than Lilly had ever "heard before."[24] Listening to the recordings of the dolphin's "whistles, buzzings, raspings, barks, and Bronx cheerlike noises," Lilly was convinced that the animal was attempting to communicate—to break through the interspecies communication barrier for perhaps the first time in history.[25]

In Lilly's written work, these two episodes (the beach in Maine and the lab) are registered almost as myths. And from these myths, he derived a basic hypothesis regarding dolphin intelligence as well as a theory about why the hypothesis has been obscured. His hypothesis was that equivalence in brain size means equivalence in cognitive capacity. Lilly repeats incessantly in his early writings that because dolphins have roughly the same sized brains as humans, they must necessarily have similar "computational" abilities.[26] His hypothesis was also a kind of wager, a bet on cetacean capacities that he justified through a strange mixture of logic, ethical imperative, and faith. In an argument whose circularity Lilly himself would likely have admitted, he wrote, "We have found that, in dealing with such a large-brained mammal, we must keep the working hypothesis in mind that 'they are highly intelligent and are just as interested in communicating with us as we are with them.' . . . If we use any other hypothesis, we have no success whatsoever in dealing communicatively with them."[27]

Although Lilly presents himself here as a studious scientist testing a hypothesis, this hypothesis (if it can meaningfully be called that) was not supported by a single shred of evidence. This explains Lilly's rhetoric, especially his insistence that the hypothesis *must* be maintained and his implication that failure to do so could lead to a missed opportunity of world-historical importance. In one instance in the mid-1960s, Lilly hired a young woman named Margaret Howe to teach dolphins how to speak at the research institute that he had founded. Lilly wrote, "One basic purpose behind this project is expressed by the 'mother' involved as 'no matter how long it takes, no matter how much work, *this dolphin is going to learn to speak English!*'"[28] He insisted that humans must have "faith" in dolphin intelligence: "This faith is in the working hypothesis that both we and they are intelligent enough to break the interspecies communication barrier between these very different minds."[29]

Lilly's declaration about the importance of faith to his work underlines his odd relationship to mainstream scientific inquiry. On the one hand, he was a bona fide scientist, garnering funding from major agencies, including the National Science Foundation, the Office of Naval Research, and NASA. (If the agencies were major, the amounts they awarded Lilly seldom were, however.) On the other hand, Lilly touted himself as an outsider who continually thumbed his nose at established conventions, unabashedly pulling together concepts like rigor and faith. He wrote that he had faith in a working hypothesis, but it would be more accurate to say that his working hypothesis was based on nothing other than faith.

If Lilly's wager was that dolphins are every bit as intelligent as humans, he also needed to explain why no line of communication has ever been opened between the two species. He attributed the challenges of human/dolphin communication to the material make-up of the two creatures, as well as to their living environments. Lilly discuses in turn (1) differences in vocal and auditory anatomy, (2) incompatibilities in the spatial and temporal scales of environment, and (3) distinct environmental communication media (aquatic and terrestrial). Significantly, Lilly viewed each of these as superficial, mechanical differences overlaying a fundamental equality of brains, or even—as the title of his 1967 book attests—of minds.[30] He embarked on a lifelong quest to develop technologies that might adequately relay information from the mind of a dolphin to the mind of a human, and vice versa. If his hypothesis was that human and dolphin brains possessed equal cognitive capabilities, much of his life's work was devoted to revealing that equality beneath or behind differences in physiology, scale, and environmental communication media.

In what follows, I address Lilly's three obstacles one by one.

First Obstacle: Anatomical Difference—Vocality and the Will to Break Through

In addition to considering brain size, Lilly asserted that the animal selected for research on interspecies communication "should be capable of making *sounds* with its vocal apparatus, either in air or in another medium, which sound like human-produced sounds."[31] On these grounds, the bottlenose dolphin (*Tursiops truncatus*) was an awkward choice. "Dolphins play pizzicato on a small violin," Lilly mused, "whereas human voices are more like organ pipes."[32] Even so, the bottlenose dolphin was suitable in other respects. For example, unlike other large-brained mammals (such as the elephant), the bottlenose dolphin is a manageable size. And Lilly, of course, had his own reasons for focusing on cetaceans, even if they often proved inconvenient.

The vast majority of dolphin vocalization happens underwater, either for communication purposes (through whistles or burst pulse sounds) or for spatial orientation (through echolocation by means of clicks). In current biosemiotic terms, whistles and burst pulses are primarily deployed as indexical signs pointing either to external referents or to internal states. Echolocation is more elaborate, but a cursory description would be something like this: dolphins and other animals such as bats use icons to map their surroundings in an exploratory manner, and they use indexes to signal where the echolocated objects are in relation to their own body.[33] The physiological mechanism of these vocalizations was only beginning to be understood in the 1960s, and even today there are important gaps in our knowledge. In broad strokes, however, marine biologists have determined that dolphins lack vocal cords and that they produce sound by means of so-called "phonic lips" (internal structures lying just beneath the air plug) that project air into nasal passages (or sacs). Pointing to this bilateral structure, Lilly determined that dolphins are "stereophonic" animals with the capacity to simultaneously produce two distinct sounds.[34]

Dolphins vocalize up to 150 kHz—far above the human hearing range of about 20 kHz, and they do this underwater, where our own auditory system hardly functions. "The physical conveyers of meaning may not overlap ours at all," Lilly lamented.[35] But he quickly determined that a region of sensory overlap does exist, and he focused his research on this narrow band. In his early experiments, Lilly analyzed dolphin phonation through a carefully designed system of tanks, hydrophones, tape recorders, and spectrograms. Individual dolphins were placed into tanks with hydrophones, and sounds picked up by the hydrophones were recorded onto tape and processed by early spectrogram devices, such as Kay Electric's "Sono-Graph," which printed out crude analyses of a sound's spectral components.[36]

If dolphins can vocalize up to 150 kHz, they also *generally* phonate at far higher registers than do humans: echolocation clicks sound well above the threshold of human hearing, and dolphin whistles are registered by humans as "squeaks" at the upper limits of our auditory range. Lilly was therefore understandably delighted—and more than a bit perplexed—when a particular dolphin named Chee Chee began vocalizing at a frequency range close to the human voice. How was this possible?[37] An analysis of Chee Chee's recorded voice revealed only two high frequencies (ostensibly one from each nasal sac) far beyond the subjective assessment of Lilly and his colleagues.[38]

But if this was so, then where were the low sounds coming from? Lilly's explanation was that the true source of Chee Chee's low frequency vocalizations seemed to be difference tones (or "'beat' frequencies") generated from the interaction of Chee Chee's stereophonic utterances. The dolphin was not directly producing the low tone; rather, she was producing two high tones resulting in the subjective effect of a lower difference tone (see fig. 1.1).[39]

Lilly provided the astonishing explanation that because Chee Chee was physically unable to produce sounds in the preferred human range, her "extensive experience with our low-frequency [human] voices" led her to develop "the best low-frequency voice that she could."[40] In other words, this dolphin—hampered by her body—contrived an ingenious method for speaking to us in our preferred range by using two high tones (the only ones she was capable of producing) in order to generate a third, subjective, and lower tone.[41] All of this struck Lilly as evidence of the dolphin's will to break through.

In addition to the two types of underwater sounds (clicks and squeaks), dolphins are able to "vocalize" by means of their blowholes above the water's surface. Such above water outputs, Lilly proffered, "are emitted in the air, apparently to accommodate to us in our medium. They lift the blowhole up in the air, open it, and make very loud sounds."[42] He insisted repeatedly that these actions were conscious attempts, on the dolphin's part, "to communicate with us as they do with one another."[43]

Because of his conviction that the dolphins were *speaking* to him

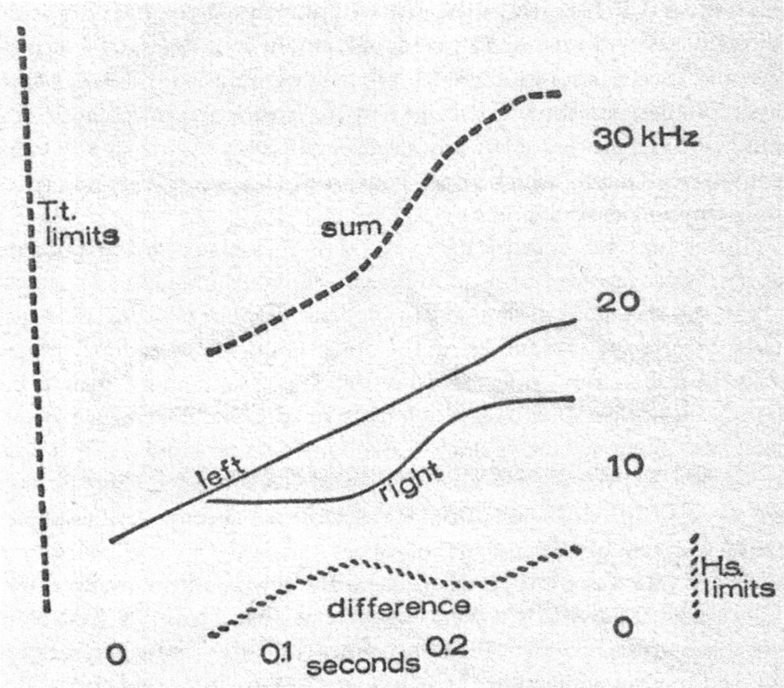

FIG. 1.1 Graph of Chee Chee's "difference tone." John Lilly, *The Mind of the Dolphin: A Nonhuman Intelligence* (New York: Avon, 1967), figure 6, p. 139.

through their blowholes, Lilly began referring to blowhole plugs as "tongues." At the Communication Research Institute, Margaret Howe capitalized on this analogy by adding a visual component to her English lessons with a dolphin they called Peter. Howe applied thick white makeup to her face and donned black lipstick to accentuate the contours of her mouth as she spoke (fig. 1.2). Peter's "eye was in [the] air looking at my mouth," Howe later recalled. "There was no question. . . . He really wanted to know: where is that noise coming from? What is the sound?"[44]

Even after long-term training, Howe never managed to elicit more than weak mimicry of a few words from any dolphin. Once again, Lilly chalked the failure up to bodily incommensurability and set about developing a more robust system of mediated translation. His project, JANUS (Joint Analogue Numerical Understanding System), was an attempt to mediate human/dolphin communication through a complex technological apparatus of codes and symbols. The idea was to train dolphins placed in tanks to recognize "dolphin-comprehensible wave forms matched to

FIG. 1.2 Margaret Howe with makeup. Still from *The Girl Who Talked with Dolphins* (film), directed by Christopher Riley (London: BBC, 2014).

dolphin viewable symbols on an underwater screen."[45] But JANUS also failed to deliver: it appeared that Lilly was simply unable to devise a method of transporting ideas from one brain to another. The dolphin's body (its vocal and auditory systems in particular) was just too different from the human's, and Lilly was at his wit's end trying to find a way to bridge the chasm. After years of failure, faith alone could sustain Lilly and his colleagues. As Antoinette Lilly, John Lilly's second wife and director of the Human Dolphin Foundation, was wont to say, "The question is not whether dolphins are intelligent enough to communicate with us, but whether we are intelligent enough to communicate with dolphins."[46]

Second Obstacle: The Problem of Scale

At times, Lilly conceptualized the problem of bodily incommensurability in terms of scale. Consider the following observations:

- "The limiting velocity of our communication of sound in air is 340 meters per second. . . . The dolphins' vocal communication has a limiting velocity in seawater of 1540 meters per second. . . . Thus their sonic communication velocity is about 4.5 times that of ours."
- The dolphin has "2.25 as many neurons [as humans], i.e., 2.25 as many pathways into his brain, each of which is operating effectively at 4.5 times the frequency."

- "[B]ecause all of his frequencies are multiplied by a factor of 4.5," a dolphin receives "four and a half times the amount of information that we [do] for each second."
- A "dolphin can put out ten times the sonic physical information per second that a man produces."[47]

Taken together, observations such as these constitute what I will call a *proportional logic*. This logic suggests that dolphins are just like humans—they simply function at different speeds and sizes than we do. It suggests that dolphins also have neurons, and they also receive information—it is just that they have more neurons than we do and thus receive more information in shorter periods of time. Understood in these terms, the dilemma of human/dolphin communication was no longer simply one of vocal and auditory mismatch. Interspecies communication was instead conceived as a *proportional* problem of translation across scale. Effective communication with dolphins would require a purely quantitative transformation based on proportion or ratio.

Technological innovations driven by the Cold War presented important opportunities for applying this proportional logic. For one thing, he made use of a vocoder to shift his voice in real time into a frequency range closer to a dolphin's. Employing a unique "human-to-dolphin coder" designed specifically by Will Munson, Lilly's voice was "stepped up in pitch by a factor of ten" before being transmitted to a dolphin via a hydrophone placed in a tank.[48] While the idea here again was that vocal and auditory differences rendered interspecies communication possible, this time Lilly was not satisfied with locating a common frequency strip—whether through a dolphin's gymnastic abilities to produce different tones (as in the case of Chee Chee) or by luring a dolphin out of water and coaxing it to "speak" (as in the case of Margaret Howe and Peter). Rather, Lilly used technology to "step" his voice up, thus catapulting his voice into the dolphin's hearing range.

Tape provided another potential solution to the problem of scale.[49] Lilly observed, "We found that most of the [dolphin] vocalizations made far more sense and their inherent complexity showed up much more easily when we extended their duration and lowered their pitch by slowing down the tape."[50] Notice the way that Lilly develops an analog relationship between humans and dolphins that positions dolphin vocalization as both *higher in pitch* and *faster* than human speech. Dolphin speech is simply human speech sped up, and magnetic tape technology provided Lilly with both the language and the conceptual resources to make this analogy.[51]

Lilly was probably not the first to make such an analogy, and he certainly was not the last. Roger Payne opens his memoir, *Among Whales,* by musing that the difficulty of studying whales is attributable to the fact that they inhabit different a "time scale." Seen from a human perspective, he says, whale *play* does not look like play at all—this is because we expect play to be quick, whereas whales play extremely slowly by our own human standards. "In order to observe whales," he says, "you must be willing to set your metronome to adagio. Then, to understand what you have seen, you must fast-forward through your observations by setting your metronome on allegro."[52] Rather than the metaphor of tape, Payne invokes the more specialized (and in some ways, more archaic) figure of the metronome. But that musical tempo-keeping device does not allow him to make the point he wishes to, which is why he must appeal to the odd notion of (re-)setting one's metronome in order to "*fast-forward* through your observations." A fascinating (mixed) metaphor, indeed!

Payne makes an additional argument that has been taken up more recently by others: if dolphins are humans sped up, birds are accelerated whales. In an online article accompanying the publication of his book, *Thousand Mile Song,* musician and ecologist David Rothenberg speeds up a humpback whale song ten times and the result sounds uncannily like a nightingale.[53] Rothenberg notes about whale song that humans "have trouble perceiving the structure by which the sounds are organized because the notes seem cast out in slow motion, with relatively long silences between each unit of sound."[54] His point, like Lilly's, is that translation media close the gap between species. Translation media (in Rothenberg's case the technology is digital) illustrate formal similarities that play themselves out on different scales.

Needless to say, Lilly never managed to speak with a dolphin by speeding up his voice—nor, for that matter, has anyone managed since. Still, he maintained until the end that, theoretically, at least, and from the vantage of his proportional logic, mammalian minds mirror each other across environmental differences and through deep evolutionary history. As he wrote about dolphins, "Here was an animal who from the viewpoint of evolutionary theory is in a group of mammals who have developed for the last thirty million years in the sea, completely separated from the evolution of primates which gave rise to *Homo sapiens.* His anatomy and physiology, though strictly mammalian, were of a strange and different form than ours, including his vocalization apparatus."[55]

Despite these tremendous differences Lilly insisted that when he came into contact with dolphins, he experienced a "feeling of weirdness"—a feeling, that is, that he and dolphins were *essentially the same.* But millions of years of evolutionary bifurcation are not so easily bridged, as Lilly's many failed experiments so poignantly testify. Despite postulating a proportional relation between man and dolphin, Lilly found no success in bridging the chasm with vocoders or with tape.

Third Obstacle: The Problem of Environmental Communication Media

Lilly was keenly aware of another difficulty in breaking through the interspecies communication barrier: dolphins live in water and humans live on land. "[I]f we are going to stay in air and speak to dolphins," he wrote, "and they are going to stay in water and speak back to us, we must create (by electronic or other means) the complete chain of communication so that they can hear what we are saying, and we can hear what they are saying with equal facility."[56] Now, this could be viewed as merely a technical problem akin to differences in human and dolphin phonation apparatuses that I have already discussed at length. But what starts as a purely technical matter began, over the course of Lilly's career, to take on metaphysical connotations. As part of his injunction that we must "meet dolphins halfway," Lilly insisted that humans must understand what it means to dwell in water. "We must learn to live wetly," he wrote, advocating an orientation to the world that he termed "wet courage."[57] Lilly believed that the only logical path forward was to immerse himself in the aquatic world of marine mammals, to inhabit the world—to some extent—as they do.

In order to do this, he set about creating a research facility at which researchers could be in contact with dolphins year-round in as natural an environment as possible. With funding from various sources, Lilly selected a plot on the island of St. Thomas in the Caribbean. Having bought the plot and cleared the land, he summoned Lieutenant Commander Donald Gaither of the United States Navy's Underwater Demolition Team to the island. The Demolition Team exploded the rock leading from the Institute down to the ocean; this created a pool where dolphins could be held in captivity, but which would be perennially refreshed by ocean water. Once the facility had been completed, Lilly was able to exit the main building and go directly down to the main pool, where he, his family, and his guests could lounge about and interact with dolphins.

Lilly's plan to learn about dolphin intelligence by spending time with the creatures in as natural an environment as possible was bolstered when a young woman working at a resort on the island visited the facility unannounced and offered her services.[58] Lilly hired the woman, Margaret Howe, already mentioned in this chapter, as a kind of research assistant (despite her lack of scientific training—she had dropped out of Tulane

University after one year and moved to the Caribbean).[59] It was decided that Howe's work, which included teaching the dolphins how to speak, would be more effective if she lived with the animals—almost as one of them. Lilly's team immediately set to work turning the facility into a space where three dolphins—Pamela, Sissy, and Peter—could live side by side with Howe. The converted house that Lilly was using as a facility was flooded with water, and Howe was provided with a slightly raised bed on which to sleep. The bed was low, so that Howe could stroke the dolphins while reclining. This new design was an extension of the original idea for the institute, which attempted to soften the boundary between land and ocean, and between land- and ocean-dwelling creatures.

This did not work perfectly—far from it. Constructing a building where humans and dolphins live side-to-side, where the proximity of terrestrial space and aquatic space allows different beings to rub shoulders, posed considerable challenges. In her account from this period, Howe notes that the depth of the water surrounding her bed (sixteen inches) was adequate for her own purposes ("In sixteen inches of water a human can maneuver, walk, work at a desk, sit in a chair, eat"), but it was too shallow for the dolphins to move in the ways they wanted to. Howe suggests that in future iterations, deeper sections should be constructed, such as a thirty-inch passageway and a six-foot deep pool.[60]

Lilly insisted that Howe interact with the dolphins as though she were their mother. ("One basic purpose behind this project is expressed by the 'mother' involved as 'no matter how long it takes, no matter how much work, *this dolphin is going to learn to speak English!*'")[61] There is, of course, a lot going on from perspective of gender roles, both in the sense that Howe was understood as somehow more nurturing than Lilly could be because of her gender, but also that her labor was considered expendable in the quest to meet his goal. On the second point, Howe is yet another in a long line of women who support "great" men—one cannot help but think of Tolstoy's wife diligently hand-copying his novels, or of Thoreau's sisters cooking for him while he supposedly lived off the land, among literally hundreds of examples. On the first point, Howe always thought of herself as a *mother* teaching a *child*, when the more obvious analogy would have been a teacher instructing a student. Lilly and Howe both underplayed the pedagogical aspect of teaching a dolphin how to speak, more often framing it as a form of child-rearing.

Lilly surmised that breaching the human/dolphin communication barrier would only be possible if every aspect of a young dolphin's life ("from food to handling of excretion") was suffused with human language.[62] One is almost afraid to enquire how one might suffuse the handling of excretion with language, even though Lilly insists on it. Of course, by invoking feeding and waste removal, Lilly emphasizes the mother-son rather than teacher-student relationship between Howe and the dolphins. But the proposal also seems to indicate just how removed Lilly was from the "child-rearing" process. His suggestions about how to treat the dolphins always sound a bit abstract, as though he knows what he wants to happen but does not know how to do it. At every point of his life, it seems, Lilly delegated such responsibilities to others—including when it came to his own children. And if Howe was a kind of mother to the dolphins, Lilly was something like an absent father, or a patriarch who controls the parameters but refuses to get his feet wet, as it were.

Howe's experiment was, unsurprisingly, an unmitigated disaster. I describe the episode in more detail in the following chapter, but for now it is enough to say that the attempt to "live wetly" all the while exhibiting "wet courage" helped not a dot. Although Lilly remained steadfast in his belief that dolphins are trying to communicate with us, his attempts to "meet dolphins halfway" (or, more accurately, to pay Margaret Howe to meet them halfway) did nothing to overcome the human/dolphin communication barrier. The experiment did, however, cause much distress for the dolphins and much consternation for Howe.

The impulse to live wetly was related to a second branch of Lilly's work that, at least initially, he seems to have conceptualized as independent from interspecies communication. In the mid-1950s, Lilly was the inventor of flotation tanks (also known as isolation tanks): enclosed water containers used for sensory deprivation. While these devices were originally understood as having a primarily military function, in the 1970s Lilly repurposed the tanks as a way to "commensurate" humans with dolphins.[63] In *The Mind of the Dolphin*, he invites us to venture "beyond our dry-featherless-biped-on-the-dry-ground philosophy" and to "go into the water to meet the dolphin."[64] But what he has in mind here is nothing like what he had attempted with Margaret Howe on St. Thomas, for even there, Howe was forced to live a primarily terrestrial life, with her body suspended partially in but not always immersed in water. By contrast, Lilly describes what it feels like to stay in a flotation tank for several hours. "For a while there is only the contrast; thinking before, in the dry, civilized life, and thinking now in the wet, civilized life. We are preparing here as if we are setting out to prepare for a new mathematics, a new logic never before seen or discovered by man."[65]

Although Lilly delegated research having to do with care-giving and parenting to a woman, he is happy—even eager—to participate in other kinds of experiments. Earlier in the chapter, I wrote about his gruesome experiments with sleeve guides. With isolation tanks, he used himself as a research subject, thereby probing the boundaries of his own consciousness. Like many male scientists of his time, he delegated certain tasks to women and laborers, but history also bequeathed to him a different scientific inheritance: from Johann Wilhelm Ritter and Alexander von Humboldt to Benjamin Franklin and John Stapp, a number of scientists have positioned themselves as courageous men willing to put their own bodies and minds on the line in the quest for knowledge.

Lilly says that wet life has a new logic unknown to terrestrial dwellers. What might this "new logic" look and sound like? Lilly does not offer a satisfactory answer to the question. Instead, his investigation brachiates, moving in two distinct directions, neither of which genuinely considers the kind of aquatic logic that he himself invokes. The first direction leads essentially to conceptualizing the dolphin as a noble savage, while the second leads Lilly away from water and toward the inner recesses of the mind. What begins as an investigation into aquatic logic ends as another repudiation of media.

Infinite Withdrawal

THE DOLPHIN AS A NOBLE SAVAGE

Lilly argued that dolphins are essentially humans who took a different evolutionary path. "May there not be other paths for large brains to take," he queried, "especially if they live immersed in some other element than air?"[66] For Lilly, the deep evolutionary history of dolphins underwater resulted in a streamlined body without arms or hands, which has also meant a lack of *tools*. A second obstacle to tool production is the ocean itself, with its strong currents and its capacity to dissolve or erode materials.[67] Lilly set himself the task of understanding this peculiar creature with tremendous intellectual capacity that had nonetheless been deprived, because of its body and aquatic living environment, of making anything of lasting value. (Of lasting value to whom? This, of course, was the kind of question that Lilly would not have been able to answer, let alone pose to himself.)

His first approximation at an answer was simply that dolphins derive the benefits of high intelligence without the destructive tendencies fostered by tools.[68] But this intuition quickly left water as an ecological environment and devolved into a naive celebration of life in the absence of production. Lilly ultimately argued that dolphins live without tools, not because of their bodily capacities or their living environments, but because they *want* to live that way.

Lilly wrote, "[W]e often ask, 'if dolphins are so intelligent, why aren't they ruling the world?' My very considered answer to this is that they may be too wise to try to rule the world."[69] Lilly was also fond of narrating the following parable:

> I wish to tell you something of what we have learned of a group of uninhibited nudists who have never worn clothes. They have never walked on their own two feet. They have no property. They cannot write their own names. They have no commerce or stores. They have no radio, no TV. They have no fireplaces, no furnaces, or any fire at all. They have no atomic or nuclear bombs, or power plants. They have no written or printed records. They have no libraries or paintings. In spite of all these handicaps, they are successful. They have big brains and bright minds. They have built-in transportation. They have readily available food supplies. They have the sense to go south in winter and to go north in summer. They have the ability to outthink, outmaneuver, and fight successfully against their enemies. Finally, they think enough of us to save each of us when they find us in trouble.[70]

These kinds of parables are of course part of a particular historical moment in which the West turned to various "Others" as a way to deal with its own political and epistemological crises.[71] In the 1970s and 80s, using dolphins for this purpose became ubiquitous. For example, science fiction author Douglas Adams mirrored the Lilly-esque position on dolphins in his bestselling *Hitchhiker's Guide to the Galaxy*: "On the planet Earth, man had always assumed that he was more intelligent than dolphins because he had achieved so much—the wheel, New York, wars and so on—whilst all the dolphins had ever done was muck about in the water having a good time. But conversely, the dolphins had always believed that they were far more intelligent than man—for precisely the same reasons."[72] For Adams, Douglas, as for Lilly, dolphins may be said to be more intelligent than humans precisely *because* they "muck about in the water having a good time" rather than inventing and producing things.

It is difficult to know how to read these statements by both Adams and Lilly—one cannot help but detect a hint of irony. That said, no matter how we choose to read these passages, it seems that neither Lilly nor Adams is speaking about dolphins any longer. It is clear that the target of this discourse is ultimately Western humanity's malaise rather than cetacean intelligence. Far from exploring a new logic, Lilly and Adams promulgate a romanticized negative projection of a society in crisis.[73] For writers like Lilly and Adams, dolphins fulfill the role of "noble savage" in the same way that "primitive peoples" did for Rousseau two centuries earlier.

THE DOLPHIN AS A MIND

In the 1970s, Lilly began to speak increasingly about the flotation tank as a device that offered him the closest possible experience of what it is like to be a dolphin. The idea was wrongheaded in my opinion. To be immersed in a flotation tank is to be immersed in water—point taken. But where in a flotation tank might one find the rich properties of the ocean as a medium: rapid sound transmission and cavernous spaces for echolocation? Oceans are resolutely social environments, at least if one is willing to be flexible with the meaning of "social," while flotation tanks are by definition isolating. The comparison between an ocean and a tank is therefore something of a stretch, to say the least. Even so, it helps us understand the terms of Lilly's project.

The aspect that ultimately interested Lilly about flotation tanks was sensory deprivation. He began to view the tanks as a way to access the mysterious inner spaces of the mind, what Sidney Cohen (quoted in Lilly's *The Center of the Cyclone*) called "the beyond within."[74] From the 1970s onward, Lilly increasingly equated external, physical space with the inner space of the mind, and he understood the limits of outer space as roughly equivalent to the deepest and most hidden parts of the psyche. "I realized, finally," he wrote in *The Deep Self: Profound Relaxation and the Tank Isolation Technique*, "that the depths of mind are as great as the depths of cosmic outer space. There are inner universes as well as outer ones."[75]

If the inner space of one's own mind and the outer space of the cosmos are ontologically similar—if the two are equally "deep," and if both harbor profound secrets—Lilly was also quick to point out that they require different modes of investigation. He noticed a similar asymmetry when it came to investigating his own consciousness versus the study of other, "alien" minds. As he had painfully learned through decades of failed dolphin experiments, examining other minds is a project riddled with difficulties. Access to one's own mind is best achieved, Lilly suggested by contrast, not through all manner of translation media, but by combining watery isolation with psychedelic drugs. Examining the "beyond within," Lilly wrote, bypasses those "self-limiting languages" so necessary for communication in "external reality": "In the inner domains the observer-agent is not limited by his/her communicative capacities."[76] In other words, investigating one's own mind in an isolation tank short-circuits communication and bypasses the vagaries of external reality.

In the flotation tank, says Lilly, the mind is shorn of its senses as well as of the material media those senses require in order to function. In certain states, the "contained mind," that is, the mind contained by the body and its sensorium, becomes, for Lilly, the "uncontained mind."[77] Isolation, in other words, is not reducible to simply being *cut off*; the experience can, in fact, lead to a "fusion between Self and one's surroundings." In the tank, says Lilly, "One no longer has boundaries, one spreads out and becomes some of the flowing materials, the flowing energies."[78]

In his later autobiography, Lilly linked the heightened external awareness made possible by extreme isolation back to interspecies communication. In a passage that is speculative even by Lilly's standards, he wondered whether dolphins had been communicating with him *while he was in the tank* by some "means unknown at present."[79] By placing himself in a tank of water and severing all sensory connection with "external reality," Lilly believed that he might have finally broken through to his dolphin companions.

In July 1970, exactly a month before Payne released *Songs of the Humpback Whale*, Lilly traveled to the Arica Institute in Chile, where he underwent training with the institute's founder, the mystic Oscar Ichazo. During that trip, as well as in the months of preparation leading up to it, Lilly kept a journal of sorts, which allows an unusually intimate glimpse into his thinking at the time.[80] I dwell on this document at length because it contains some intriguing statements that will help bring us full circle to whale song. A fuller elaboration would place the document in its wider cultural context, which I have done only summarily in this chapter. Lilly traveled to Chile at a moment when interest in mysticism and Eastern philosophy was high in the United States, affecting everything from politics to art. Psychedelics were also in vogue (Lilly would later go on to befriend Timothy Leary).

In the opening pages of the journal, Lilly writes repeatedly, if elliptically, about "ego and illusion," about the "non-importance of ego," and about getting "rid of ego." He repeatedly mentions water and music as linked mechanisms for eradicating the pernicious ego. "Music = cleaning by water" he writes, and then on the following line: "Clean the mind of associations." "Ego can't stop the rhythm of bolero," he scrawls on the next page, after mentioning Ravel and "bolero 'inspired' music."[81]

On July 6, 1970, Lilly recounts a dream from the previous night and then offers an interpretation. In the dream, he is out at sea when he real-

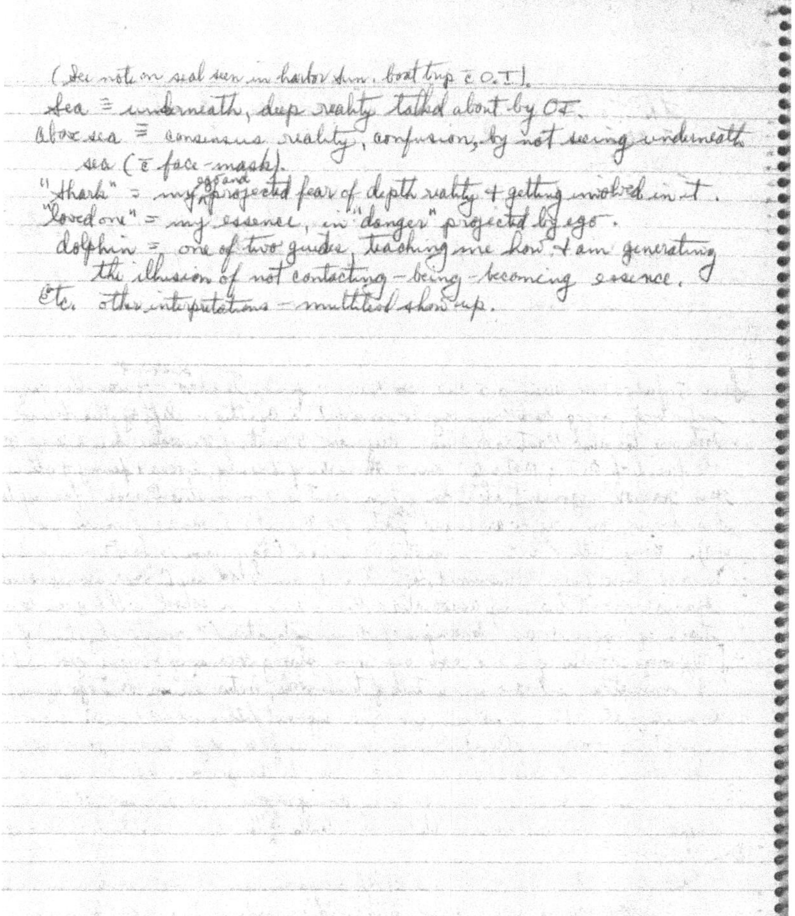

(See note on seal seen in harbor dream, boat trip c̄ O.I.)
Sea ≡ underneath, deep reality talked about by O.I.
above sea ≡ conscious reality, confusion, by not seeing underneath sea (c̄ face-mask).
"Shark" = my expand projected fear of depth reality + getting involved in it.
"loved one" = my essence, in "danger" projected by ego.
dolphin = one of two guides, teaching me how I am generating the illusion of not contacting—being—becoming essence.
Etc. other interpretations—multilevel show up.

FIG. 1.3 Lilly's dream analysis. Courtesy of the Department of Special Collections, Stanford University Libraries. M0786, John Lilly Papers, box 52, folder 8.

izes that someone he loves is under the water. He sees a shark fin and believes that his loved one has been devoured, but then realizes that all is well because the creature below the surface is in fact a dolphin. His interpretation decodes the constituent elements (see fig. 1.3).

Here is a transcription of Figure 1.3:

Sea = underneath, deep reality talked about by O. I. [i.e., Oscar Ichazo]
above sea = conscious reality, confusion, by not seeing underneath sea
(w/ face-mask)

"shark" = my ego and projected fear of depth reality + getting involved in it
"loved one" = my essence, in "danger" projected by ego
dolphin = one of two guides teaching me how I am generating the illusion of not contacting-being-becoming essence

Toward the end of the journal, on August 10, Lilly begins teaching himself basic music theory (see fig. 1.4) and then writes that "everything becomes music." In the journal, music and water perform the same work.

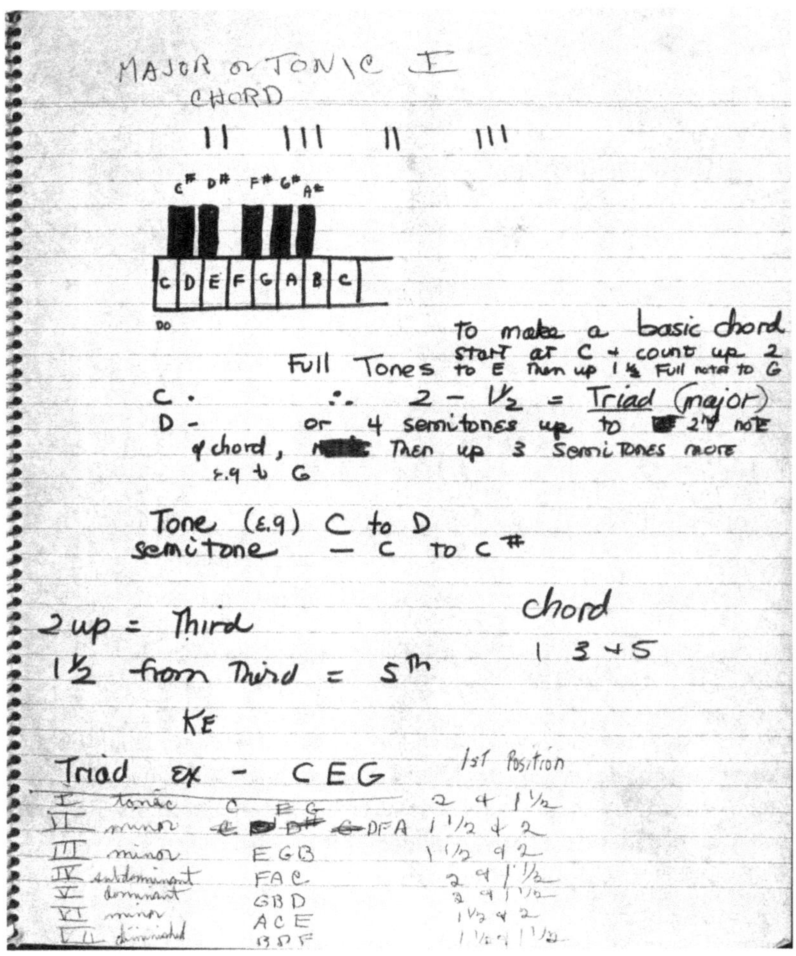

FIG. 1.4 Lilly teaches himself music theory. Courtesy of the Department of Special Collections, Stanford University Libraries. M0786, John Lilly Papers, box 52, folder 8.

Both are immersive and ego-destroying. Both establish communication without mediation.

Considering the scarcity of music in Lilly's published work, it may seem surprising to find music so prominently foregrounded in his journal from 1970. But Lilly would continue to mention music sporadically in his later works, and it makes some stunning appearances in his pre-1970 writings as well. In his 1972 autobiography, for example, Lilly wrote about his first LSD trip a decade earlier—an experience that he treasured, and to which he credited many of his subsequent ideas. During the trip, Lilly recalled, he lay down and "went with Beethoven's *Ninth Symphony*." After the music had "entered" him and "programmed" him, Lilly recounts moving "with the music to Heaven" and seeing "God on a tall throne." In that state, Lilly registered the chorus in the fourth movement as "angels praising God."[82]

Another curious passage about music appears in *The Mind of the Dolphin* (1967). If a dolphin's brain and therefore mental capacity is equal to our own, Lilly muses, how might we even begin to describe the monstrous intelligence of the whale? Lilly asks the reader to imagine what it would be like to be a sperm whale, able to "relive a full playing of Mahler's *Messiah* or any of your favorite symphonies without any apparatus or recordings outside of your own brain!"[83] He continues: "This huge brain of the sperm whale may also have re-programming abilities for the symphony. He may modify the music and even further elaborate it beyond any human conceptions of music."[84]

These two examples elucidate his understanding of music as something that happens outside of material media, either in the spiritual realm or in the disembodied computational brain—which is to say, in the mind. In his journal, music is a "cleaning by water" that washes "the mind of associations." In his 1972 autobiography, music transports him directly to an angelic choir in heaven; and in one of his most popular published works, *The Mind of the Dolphin*, the incredible intelligence of the whale is illustrated by its alleged ability to play a full symphony in its brain, without any material apparatus.

Water/Music

It would not be an exaggeration to say that Lilly spent his entire life railing against media. He assumed the equivalence of minds in large-brained mammals, and he demanded a seamless communication between those minds. In the 1960s, Lilly tackled the problem of mediation with more mediation. Responding to what he viewed as the obstacles between human and dolphin communication, he tried all manner of strategies and techniques: using hydrophones, tanks, and vocoders; building new research facilities; and employing people to teach dolphins to speak. Especially during the 1960s, Lilly seemed to believe that no problem caused by media was too great to be solved by other, better media. A human's phonation apparatus won't let us speak dolphinese? No worries, Lilly will create a translation device to overcome the problem. We cannot communicate properly with dolphins because we do not live in water? Have no fear, because Lilly can draw upon the resources of multiple US funding entities to create a bridge between aquatic and terrestrial life.

But no amount of intervention worked. All the hydrophones in the world were not enough to break through the "interspecies communication barrier." When communication was not forthcoming, and when multiplying the layers of mediation did nothing to attenuate the problem, Lilly became increasingly frustrated with media rather than increasingly doubtful of his hypothesis regarding dolphin intelligence. When, at the end of the 1960s, he finally gave up creating a conveyer belt of meaning between the human and dolphin mind, he turned dramatically to transcendental forms as a way to bypass mediation *in toto*. If sound was a stubbornly material entity in Lilly's books, subject to the vagaries of its propagation medium, then water and music were ways to establish connection without mediation. This is a textbook example of what Antonia Majaca and Luciana Parisi refer to as an apparatus that is "fixed and ossified in [its] impossible desire to escape," and in its "continuous search for a transcendental realm 'beyond' the instrumental."[85]

European thought has often harnessed music and water as dual ideological ways to transcend the instrumental. About music, Ochoa Gautier writes, "To culture's increasing enmeshment in 'the immaterialization of capital' [beginning in the 1970s] correspond other forms of immaterialization and deracination as well, and sound/music is particularly suited for such immaterialization precisely because of the historical difficulty of grasping its 'object.'"[86] It is also possible to detect a much older ideological configuration at work. Roger Grant has shown how music theorists in the eighteenth century deployed music as a way to bypass traditional structures of mimesis and representation.[87] Johann Nikolaus Forkel (1749–1818), for example, wrote that the relationship between musical tones and human emotion is "*ganz unmittelbare* [completely unmediated]."[88] Grant observes a long-standing tradition in aesthetic theory (beginning in the eighteenth century but renewing itself periodically, including in the second half of the twentieth century) in which music offers itself as a "form of mediation for free."[89]

Water, for its part, has been subject to a particular form of valorization in late twentieth-century theory and art practice. As an example, consider the following passage from Joan McIntyre's seminal volume, *Mind in the Waters*: "When a human enters the water, what becomes apparent is the intimate connection between mind and body that the sea forces on her creatures. Without the alienating presence of objects and equipment, with only the naked body encasing the floating mind, the two, split by technological culture, are one again. The mind enters a new modality, where time, weight, and one's self are experienced holistically."[90]

McIntyre's description is more ideologically motivated than empirically grounded—that goes without saying. But her claim is also symptomatic of a larger trend. Since the 1970s, water has frequently been invoked (and used) as a medium of spiritual healing by New Age practitioners. The ocean in particular is often depicted as a setting for the unbounded and the natural sublime.[91] From the perspective of space, Earth is not registered primarily as a habitable landmass but as a "blue marble." Water is the image of Earth as spiritual center and home; like music, it is believed by many to create holism.

Stefan Helmreich's critique of underwater music composition brings the water/music dyad into sharper relief. He focuses on a word that joins the two terms in the dyad: *immersion*. Water and music are frequently considered "media" in which one can be immersed—no boundaries or jagged edges, just pure immersion. As an example, Helmreich considers a piece by French composer Michel Redolfi that was performed in a pool at Dartmouth College in 1982. Redolfi's own description of the piece makes the water/music parallel explicit. "Listeners of the Underwater concerts 'immerse themselves not only in the 90-degree heated swimming pool, but in the sound itself.'" But as Helmreich observes, the production of Redolfi's underwater music required an elaborate system of transduction and mediation, including microphones, speakers, and a Sonic Waters Safety System (SWSS). "As this list of equipment makes clear," writes Helmreich, "'immersion' is a hard-won pun, a hard-won laboratory effect. . . . The *symbolism* of water as a sublimely immersive medium must be actively realized technologically."[92] The point, for Helmreich, is that the "immersive" character of both water and music is in fact densely mediated. "Immersiveness" and "connection without mediation" are illusory *effects* produced by careful technological manipulation.

As this section has shown, there is a consistent appeal to music and water as figures that either transcend or simply bypass mediation. We see this in popular culture today, in colloquial language surrounding all kinds of performance and aesthetic experience—listeners often report something like an oceanic feeling, of being awash in sound. For an additional historical example, recall Freud's *Civilization and Its Discontents*, which opens with a message from a correspondent who describes the overwhelming sensation of subjective dissolution in religious rites as "oceanic" (*Ozeanischem*). The unnamed correspondent was Romain Rolland, winner of a Nobel Prize and one of the most important early twentieth-century writers on Beethoven (and other musicians).[93] In his deployment of music and water as ideological figures, John Lilly exemplifies the appeal to a kind of oceanic thinking.

Reconceptualizing cetaceans in the 1960s and 1970s was also spurred by the notion of music as a universal language. The idea of a universal language is oxymoronic, since languages are always particular to a culture or society. But, at least in the realm of discourse surrounding cetaceans, the oxymoron of music as a universal language proved potent. Associated with this metaphysical notion, whale song became a form of aesthetic production that anyone, anywhere could enjoy and appreciate.

✷

It might be useful to step back a little and take a wider look at the question of whale song. In discourse about cetaceans since the 1970s, music works in two overlapping ways. First, as described in this chapter, it works as a kind of metaphysical force allowing for frictionless connection or even communication between different beings and environments.[94] Rothenberg's commentary on his comparison between bird and whale song, mentioned near the end of the "Second Obstacle" section above, is a good example. Rothenberg writes,

> Now you hear an even blend of nightingale and whale, their actual speeds, their slowed down, sped up speeds, a kind of mélange of the whole aesthetics of songs in the natural world. Very extreme songs, that somehow all seem to fit together. There's a kind of aesthetic of nature. Here it is, singing all around us.
>
> So what does it all mean? That in evolution, some aspect of the beautiful survives. Not only what is expedient, necessary, and efficient.[95]

In this passage, the notion of the beautiful is harnessed as a metaphysical tool to bring different species into closer proximity. In anti-whaling activism, aesthetics is often deployed in a similar way. Reflecting on *Songs of the Humpback Whale* on the album's fiftieth anniversary, Roger Payne recalls that when he has played the recording for audiences over the years, people often go into a "trance." "That was the clue," he says, "that 'this is changing the lives of these people.' And that's how I think we've got it to really make a difference." The first time that Payne himself heard whale song, he was moved to tears. And he knew then, immediately, that he may well have found the secret to stopping the whaling industry.[96]

I said above that music works in *two* ways. The second way is more prosaic, having to do with a close analysis of how whale vocalizations are structured. Note well: in the early 1970s, a group of scientists (especially Roger Payne, Scott McVay, and Katharine Boynton Payne) "discovered" that, in addition to simple calls, humpback whales vocalize in complex, lengthy, structured patterns.[97] These complex patterns—which they called songs—indicate significant cognitive complexity. Along with other factors, scientific whale research in the 1970s transformed the way people think about whales.

And so, to reiterate, whale song has two distinct, if occasionally overlapping, aspects: (1) the metaphysical, and (2) the more prosaic (if not less profound) discovery of patterned vocalization. It is *not*, however, patterned structure that most listeners respond to when they listen to recordings of so-called whale songs. If, in Payne's and McVay's original formulation, whale song refers to the structure of repeated vocal units, this structure is barely audible to listeners without technological manipulation. Payne and McVay observe that, largely due to how slowly the patterns are repeated, "On first hearing humpback vocalizations, one has the impression of an almost endless variety of sounds." The songlike patterns of humpback vocalization became apparent to Payne and McVay only after printing out and carefully arranging spectrographs: in other words, by looking at the "songs" as images frozen in time (see fig. 1.5).

Only after printing and visual analysis did Payne and McVay realize that "all prolonged vocalizations occur in long, fixed sequences and are repeated with considerable accuracy every few minutes."[98] Whale song is so slow that it hardly registers in human ears as song at all: most people never hear the structure, and those that do hear it only do so through careful, repeated listening. (If you've not done so already, now would be a good time to listen to some whale song recordings! Dozens are freely available on YouTube and elsewhere.)

What people have responded to so profoundly in whale song recordings is the ethereal timbre of individual vocal units. To most human listeners, each individual call (to use a term from current scientific inquiry)

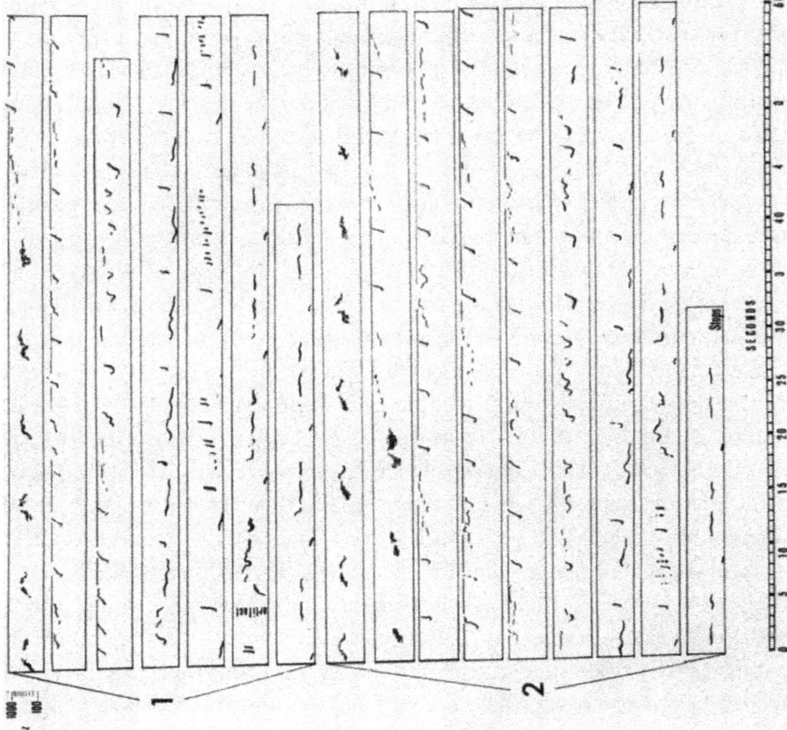

FIG. 1.5 Whale song. From Roger Payne and Scott McVay, "The Songs of Humpback Whales," *Science* 173.3997 (1971): 585–97 (figure 5, 590). Reprinted with permission from AAAS.

is sonically rich and emotionally powerful. As one author commented about whale song recordings, "these sounds came to be called songs not only because of structure, but also because of their particular sonic qualities."[99]

To put all of this in simple terms: "whale song" may refer to the structured pattern of whale vocalizations, but these are not easily audible by human listeners, who are instead drawn to the immediate (that is, nonstructural) sonic effect of whale vocalization. Although the structure of whale songs is seldom experienced directly by human listeners, such listeners are sometimes aware that scientists have "proved" that whales are intelligent creatures who sing. This knowledge partially shapes the framework for the listener's experience.

Most people I have spoken with while writing this book, from musicians and musicologists to scientists and many others, have *some* sense that whales are purported to be singers, and trust that the scientific community's assessment is correct. Many people with whom I've spoken about my book have heard whale song recordings, and have had a strong reaction, although not always a good one. Listeners' reactions to whale song recordings are based on a strange confluence of (1) the popular reception of scientific research on the actual workings of whale song; and (2) the sonic effect of individual sounds on the recordings. Both of these factors are bolstered by the ideology of musical metaphysics, which was grafted onto whale vocalization by the people discussed in this chapter.

A more critical take would go as follows. Music may *generate* metaphysical ideas or experiences, and it may produce "metaphysics" as an effect, but as a system, music is anything but metaphysical. Music is a tightly structured material-semiotic system. (For more on this argument, see the appendix.) Water, likewise, is anything but metaphysical. Indeed, to take a metaphysical perspective on water at the current moment of deep ecological crisis would be a grave error.

Viewed from a critical-historical perspective, then, it is necessary to demystify (as it were) both music and water. I even suggest that we retire the word *song* in relation to animal vocality. The notion of whale song was accurate enough for the 1970s, and at that particular historical juncture, the word did tremendous political work: it helped anti-whaling activists achieve the surprising feat of, essentially, saving the whales. But the usefulness of the word has run its course. The problem is not so much that the word is wrong (although an argument could be made for this as well) but that it is too freighted with metaphysics to be helpful any longer.

Here, I add a short but important technical observation. In their seminal paper, Payne and McVay do not use the terms *note* or *pitch* when describing whale song—instead, they say that groups of *units* form a *phrase*, groups of *phrases* form a *theme*, groups of *themes* form a *song*, and groups of *songs* form a *song session*. Let's look closely at the terms they choose: in one case they use non-musical terminology (unit), in some cases they borrow language from human music-making (phrase, theme, song), and in one case they use musical language in a peculiar, if not exactly erroneous, way (e.g., "song session"). From where do Payne and McVay derive their idiosyncratic and mixed terminology?

In the paper, Payne and McVay comment upon the contested meanings of the word *song* when applied to animal vocalization. After a substantial discussion, they claim to settle on a definition offered by W. B. Broughton: "a series of notes, generally of more than one type, uttered in succession and so related as to form a recognizable sequence or pattern in time."[100] But here is the catch: although Payne and McVay claim to use this definition, they actually do not. As I mentioned, when they describe whale song, they do not use the word "note" (as Broughton does), but instead refer to "units" as the basic building blocks. Even according to Payne's and McVay's own understanding, whale song is not a series of notes but a series of what they call "units."

A note is a discrete entity—it is the same as a pitch, to use more technical language. The units that Payne and McVay refer to, by contrast, are not individual notes or pitches but what I would call gesture-calls.[101] (See the appendix for details.) Units (or gesture-calls) are more than discrete pitches, and in this sense differ from a simple middle C. They have specific characteristics and shapes and typically consist of more than one note: a slow slide from one pitch to another, for example, or a quick succession of pitches within a narrow frequency band (corresponding roughly to *glissando* and *tremolo*, respectively, in musical terminology). You can clearly see the shapes of these units in figure 1.5. And keep in mind that gesture-calls/units are characterized not only by pitch but also by other sonic parameters: for example, a *slow* slide (that may get louder or softer over its duration). Because they have a recognizable contour and shape, gesture-calls are indeed units of a kind.

A definition of whale song should read: "a series of units [not notes!], generally of more than one type." This, as I've mentioned, is the definition that Payne and McVay actually end up adopting. But they have kept the word *note* in the definition they borrow from Broughton, because to replace it with *unit* would damage their argument that whale vocalizations are songs (human songs, after all, are a "series of *notes*"). After providing Broughton's definition, Payne and McVay write, "By that [i.e., Broughton's] definition, . . . we feel justified in using the term 'song' to describe repetitive sound patterns of humpback whales." Would they have felt equally justified had they been honest about the definition they used? I am not convinced. Change "notes" to "units," and the definition starts to sound a lot less like what we recognize as song.

So this is another good reason to stop calling it whale song.[102] Authors work themselves into corners, shifting around words (without acknowledging it) in ways that can only be sorted out with careful reading. Indeed, some readers may have been tempted to skim over the preceding close reading since it *is* tricky and pedantic. But the consequence of not doing that kind of careful work is that we end up with a conceptual apparatus that is sloppy, misleading, and inaccurate.

Last Thoughts on Lilly's Wager

By way of conclusion, let us return briefly to Lilly's experiments. We can safely dispel most of the outlandish ideas that his experiments were based on, and we ought to be skeptical of the deliberate mystifications of both music and water in his work. But what if we bring to the surface, as it were, the one aspect of communication that Lilly sought to constantly submerge? What Lilly's work shows, above all else, and despite his continual attempts to prove otherwise, is that media are integral to communication itself. Communication does not happen independently of media; rather, it happens in, with, and through various kinds of mediation. When Lilly struggles in vain to find the right kind of mediation between the human mind and the dolphin mind, what he does not understand or cannot understand is that those mediations are constitutive of communication itself.

Since Lilly's time, researchers have made important strides toward understanding marine mammal communication. Some of that work takes media into account, as evidenced by a recent article on reverberation in whale vocalization. If you listen even briefly to a whale "song" recording, you'll notice immediately that reverb is a salient acoustic element. The reverb is produced by the sonic environment of oceans: some whale vocalizations can be heard by other whales up to one thousand miles away, and along the way a song like that can reflect off all kinds of surfaces, especially dense vegetation. Until recently, researchers assumed that the reverb was a mere effect of whale vocalization, but now, for the first time, at least one scientist seems to have shown that whales take the reverberant soundscape of the ocean into consideration when they vocalize. Whales do this, observes cognitive psychologist Eduardo Mercado III, by avoiding "spectral overlap between [vocalized] units and ongoing reverberation."[103] (In this sense, whales are like singers in medieval churches—except that an ocean is just a little bigger than a church!)

Consider theorist John Durham Peters's speculative observation about dolphins.[104] Peters begins with two assumptions: first, dolphins communicate with each other in some sense—that is, they participate in semiosis even if not of the symbolic variety. Second, water and air have different properties as media—for example, sound travels faster and farther in water than it does in air. Now, despite the fact that sound travels *faster* through water than through air, the delay between sender and receiver may be extensive, since dolphins vocalize over long distances. From this, Peters infers that the moment at which a particular dolphin transmits information is perhaps less important than the location of various dolphin-listeners in the surrounding waters.[105] If we consider several dolphins distributed at different points in a particular region of the ocean, and if we imagine that dolphin A transmits information to dolphins B and C, it is possible that a *response* from dolphin B will arrive at dolphin C before A's initial vocalization. According to Peters, dolphin communication would not therefore abide by the rules of typical human conversation, in which linear turn-taking is a crucial structural and interpretive parameter. Furthermore, because "distant cetacean correspondents would not share the same exact sequence of messages, they cannot presuppose ordinal logic. For distant underwater communications with many senders, no single 'now' can serve as the fulcrum of turn-taking."[106]

The work of authors like Mercado and Peters shows that the ocean is a tremendous and fascinating medium for communication. Lilly was correct, in a way, to believe in something like an aquatic "logic never before seen or discovered by man." With such thoughts in mind, might we not now return anew to Lilly's challenge to "live wetly"? Might we not rethink and affirm a new kind of wet courage, a kind that Lilly would not dare seriously consider? This could mean many things, but one might be this: the courage to take seriously the characteristics of water as a medium of communication at a moment when the problem of water is far larger than communication alone.

As I write this, flooding in Durban, South Africa, has killed forty-five people in one day. The water level is everywhere on the rise. We are in trouble, and this trouble compels us to recognize that water is not an abstraction, an idea, or an ideology. And the same could be said of communication, as Peters arguers: the dilemma of communication is "a political problem of access and opportunity, not a psychological or semantic one of purifying the media."[107]

Lilly seemed not to understand any of this. Or, more precisely, he *disavowed* the realities of water and marine life. But the point of this chapter is not to argue that Lilly succumbed to certain ideological blind spots. My argument, rather, is that it was Lilly's de-materialized conceptualization of music and water that ultimately shaped anti-whaling activism and led to our contemporary, mystified notion of marine mammals. To some extent, that anti-media attitude came about as a response to his many failed attempts to transcend the instrumental. But we can see that attitude already even in his early experiments. Lilly attempted to bridge the dolphin mind and the human mind without any loss whatsoever. His was a dream of perfect translation, of unmediated connection.

Although Lilly wrote obsessively about speech, his real interest was always in a form of communication that transcended everything that speech entails, from speech organs (mouths, tongues, nasal sacs) and auditory organs to the media through which sound must travel to be comprehended. Music, on the other hand, played a much more equivocal role in his thinking, as this chapter has shown. Lilly never once suggested that dolphins sing, and he was not especially interested in the phenomenon of whale song either. Even so, his ideas about cetaceans mystified both dolphins and whales, and this mystification, in turn, both bolstered and was bolstered by Euro-American conceptions of music as immaterial or ineffable. As we see from his diaries and other writings from the 1970s, Lilly himself viewed music as a way to get around media, to short-circuit and bypass it.

As the examples by Mercado and Peters demonstrate, authors today have overcome some of the most problematic aspects of the thinking prevalent around Lilly's time. But a mystified notion of dolphins and whales still wafts through the popular imagination, finding a place in film, literature, tourism (i.e., swimming with dolphins), and New Age healing—to list a motley assortment of domains. Furthermore, even the most hardnosed researchers continue to refer to patterned whale vocalization as song; this does not always negatively impact research, of course, but it does create terminological havoc in many cases, hampering the advancement of knowledge, and creating barriers for meaningful interdisciplinary conversation.

One final riddle: no one knows why whales vocalize in patterns. No one knows, that is, why they "sing." Despite evidence to the contrary, dozens of researchers have promulgated the hypothesis that male whales sing to attract female mates. Again, there is no evidence for this theory and there is mounting evidence to the contrary. And still, people continue to espouse it. A certain assumption about animal sexuality (bolstered by a vulgar evolutionism) seems to underpin the argument—what else might a male humpback whale be trying to communicate other than (as Roger Payne himself recently suggested): "Hey, I'm a handsome dude ready to mate. Where are you, baby?" Payne admits to one other likely possibility, again ventriloquizing a whale: "Stay back buster or I'll bust your ass!"[108] (It would be easy, in my view, to think of other possibilities, for example, "There's loads of krill here!").

As dubious as alpha-male-based explanations are, they reveal a lot about how we think about animal behavior. Sex, intimacy, and love shape so many of our experiences with animals. It is to those issues that I now turn.

2
Tales of Love

Now let me tell you a different story. In October 2015, I traveled to Hermanus, South Africa—smack dab in the middle of the town's annual Whale Festival, a week-long eco-tourism event focused on whale watching and conservation. Both real and fictional whales had drawn me there, and I was eager to explore the resonances between the two, as well as the whale-centered humans I had been thinking about. The initial motivation for my trip came from the 2005 novel *The Whale Caller* by Zakes Mda (b. 1948), one of South Africa's most renowned authors. Set in Hermanus, Mda's novel narrates the story of a man who develops an intimate relationship with a southern right whale he calls Sharisha.[1] Because of the different worlds they inhabit—land versus ocean—the Whale Caller can only interact with Sharisha through sound. He plays music to and for Sharisha with a kelp horn, an ancient instrument fashioned from plants nourished by the ocean.

The kelp horn used by the Whale Caller in Mda's novel is an actual instrument with a long history. It is mentioned, for example, by the Swedish naturalist C. P. Thunberg in 1772 and described in greater detail by Percival Kirby in his seminal (if ideologically dubious) *Musical Instruments of the Native Races of South Africa* in 1934. Kirby mentions that the curved version of the instrument produces exactly three notes and has a "powerful horn-tone quality."[2]

In the novel, Mda tells us that when the Whale Caller plays his horn for Sharisha, she responds "with her own love calls."[3] As he plays, he dances, and to this Sharisha responds with a kind of dance of her own, a sequence of "rocking, breaching and lobtailing."[4] The instrument's sound wavers between a military bugle call and the "fast-paced scatting of a demented jazz singer."[5] Mda writes that the Whale Caller's horn "penetrated deep into every aperture of the whale's body, as if in search of a soul in the midst of all the blubber."[6] The encounter between man and whale continues throughout the night, ending only when the sun rises the following morning. The Whale Caller finally collapses in exhaustion, and Mda notes that "The front and seat of his tuxedo pants were wet and sticky from the seed of life."[7]

The Whale Caller is a strange book, which, like Mda's others, seems to teeter between allegory, Magical Realism, and deep African spiritualism. Reading Mda is like inhabiting the liminal space between wakefulness and sleep, where vivid images are bathed in an ether of unreality.

But reality surfaces at critical junctures. Throughout the novel, Mda insists on a sharp distinction between his protagonist, the Whale Caller, and the so-called whale *crier* of Hermanus, who is based on, and in fact shares a name with, the real-life person who holds this position. Employed by the town's local tourist bureau, the whale crier has a role similar to town criers of old, except that he only ever proclaims one thing: the presence of whales in the bay. "The Whale Caller was not in competition with the whale crier," writes Mda. "The Whale Caller was not a whale crier but a whale caller! . . . The whale crier alerted people to the whereabouts of whales, whereas the Whale Caller called whales to himself."[8]

In other words, the Whale Caller establishes a human/animal relationship, while the whale crier participates only in human/human relationships, and Mda takes the latter to be far more mundane.[9] Some way into the book, the Whale Caller hears the sound of the whale crier's horn. Mda mentions the crier by name: "The Whale Caller knows immediately that it is the official whale crier of Hermanus, Mr. Wilson Salukazana, the gracious gentleman from Zwelihle Township. He is alerting tourists to the presence and location of the whales."[10]

At this moment, the Whale Caller meets his mirror opposite: the mundane whale crier of Hermanus, who calls only people and never whales. But the relationship between the two is more complicated than this, because the real-life Mr. Wilson Salukazana was a kind of negative inspiration for Mda when he was writing the novel. In an interview with the journalist Angelique Serrao, Mda recounts that he first conceived the idea for *The Whale Caller* after he saw Salukazana on a television program about whales: "From the programme I thought that the man called the whales to him on his kelp horn, and I thought this was a magical thing. So I went to Hermanus to find out about this and I found that he doesn't call the whales, he lets tourists know where they are. I was very disappointed, but I thought, 'let me create this magical thing in my novel.'"[11]

What would have given Mda this false impression? What led the novelist to believe that "the man called the whales to him on his kelp horn"? And who is this entirely disappointing whale crier? Surely Wilson Salukazana's existence is not reducible to the negative image of Mda's fictitious Whale Caller? What might the whale crier think about the Whale Caller? Given the demographics of the region (the largest population group is Xhosa-speaking), I also wondered whether whales have a place in Xhosa cosmology and, more generally, what that cosmology says about human/animal relations. Both Mda and Salukazana came from Xhosa-speaking families, and the two men would have been raised with at least some of the same traditions.[12]

These were some of the issues I intended to investigate when I traveled to Hermanus in October 2015. During this period, I had been ruminating on other instances of human/animal intimacy, such as the famous example of Alex the parrot. And I had not forgotten, as I drove the winding mountain paths along the ocean, that Mda's book is really about one thing: love. Mda imagines the relationship between a human and a whale as both loving and downright sexual. The Whale Caller directs all his emotional and libidinal energy at Sharisha, who seems to offer him something that human society cannot. Love came up time and time again as I followed the threads between Mda's novel and the real people of Hermanus—even and especially when I least expected it. This chapter explores the intricate interactions between humans and whales in Hermanus before turning to examples from other times and places. It asks a few related questions: What are the ways to love a non-human? What kind of material or earthly form might this love take? Is there an ethics of human/animal love? Is there a politics?

The trip to Hermanus in 2015 was my first visit to the area since 1997. I grew up in Johannesburg (South Africa's largest city) and during my high school years twice attended a "socialist" Jewish summer camp in Onrus, a town roughly five miles from Hermanus.[13] It was strange to be back in this part of the world after nearly twenty years; it was stranger still that in some ways nothing had changed. Like many such towns in the Western Cape, Hermanus is naturally beautiful and deeply segregated. *At the cafes and restaurants that line the grassy area near the craggy cliff overlooking the bay, Black waiters from surrounding townships serve tourists and locals, including the many White Afrikaans teenagers who fill the air with laughter and plumes of cigarette smoke. Children play in rock pools far below the cliffs, while elderly Black men hawk beads and clay sculptures to passing tourists. People perch on rocks, gazing out beyond the bay toward the horizon—a diverse group of Xhosa and Afrikaans locals, visitors from surrounding areas, and a smattering of international tourists.*

I remember the scene like it was yesterday: Everyone is waiting to see a whale breach, to appear from the depths, seeming to banish time as it hovers in the air for that one jewel-like moment. A Bob Marley song seems to hover on the breeze from a restaurant whose location I cannot determine. A White girl, no older than eleven or twelve, scrunches her nose and asks no one in particular, "Wat is daardie slegte musiek?" (What is that horrible music?) I chuckle softly.

Exploring the town on my first morning there, I could find no trace of Wilson Salukazana, the whale crier mentioned in Mda's book. It turned out that a spry young man, Eric Davalala, had taken his place. As I would soon learn, Eric had assumed the mantle of official whale crier in 2011, and by late 2015 he was being touted as the world's only Whale Crier™—the position had been trademarked by the tourist bureau.[14] On the day I met Eric, the weather was gloomy, and the sightings had been nil. I opened with a leading question: "So I see you posing with this horn. But I didn't hear you blow it yet. Why?" His response came without hesitation: "OK, this is my kelp horn. I blow the horn *only* if there's whales in the bay. Most of the people thought that the whale crier blows the horn to call the whales. But he only blows the horns to *inform* the people that there are whales to be seen in the bay."[15] I was stunned. After all, Mda had said in the interview that he learned that the whale crier "doesn't call the whales, he lets tourists know where they are." Now Eric had just told me that "most of the people thought that the whale crier blows the horn to call the whales." *Most* people thought that? Which people was he referring to?

Eric clarified that by "most people" he meant Xhosa-speaking locals, not tourists. The locals' mistaken belief that whale criers communicate with whales is based, Eric suggested, on a kind of spurious causality. He told me that when he blows the horn, the whales "do something like a performance, sometimes you see them playing, or coming up and down all the time. And that's when they *thought* that now I'm really calling them." But despite his strong position regarding the ignorance of "most people," the young whale crier declared that the whales do listen. "They do hear me," he said. "Sometimes they check out—we call that spy hopping. You can see that on the poster there," he said, pointing to an educational poster on the wall inside the tourist kiosk. "The whales always want to know what is going on. It's not that we call them," he said. "They're just curious. They're so inquisitive. They like people so much."

But as he made this last comment, Eric seemed to feel that he had allowed his enthusiasm to range too far. His anxiety stemmed in part from a need to distance himself from his predecessor, Wilson Salukazana, that feckless whale crier whom Zakes Mda so sharply contrasts with the Whale Caller in his novel. I was astonished to hear Eric's thoughts: "As I said to you, most of the locals, those are the ones that have grown up thinking that Wilson Salukazana blows the horn to call the whales. Because when they come down here, they say: 'This guy [i.e., Eric] is not calling the whales anymore. Wilson Salukazana used to *call* them.' So I need to *tell* them. You see, when I'm there, I explain to them what is going on. Because they don't know."

Eric has a dual motivation for explaining things to the locals. First, it is simply part of his job to educate people about whales and their environment. But a second reason is that his own practice, which consists primarily of notifying tourists where whales can be seen, is banal in comparison with what "most of the locals" believe that Wilson used to do. Here, Mda's Whale Caller/whale crier distinction is played out again, although this time it is Wilson Salukazana (identified by Mda as a mere "whale crier") who is designated as a Whale Caller. (Or, more precisely, he is designated as an *alleged* Whale Caller, since Eric does not believe that one can in fact call whales.)

I had heard and read a fair amount about Wilson Salukazana by this point, but it was clear that I would need to meet the man in person. Finding the retired whale crier proved challenging, however. My inquiries sent me on a wild goose chase from tourist offices to museums, zig-zagging between town, harbor, and beachfront. For days, I searched in vain. And then, when I had exhausted all official channels and was feeling defeated, I turned to an approach that had served me well in earlier fieldwork in African towns and cities: I drove to the local petrol station and asked the first person I encountered. In a turn of events that seemed miraculous at the time, a petrol attendant named Philemon told me he was related to Wilson Salukazana through marriage and that he would be happy to take me to Wilson's house in the township just outside Hermanus.

In a rental car on the way to Zwelihle township, Philemon told me that Wilson was old and had poor eyesight, but that he was well looked after by his children, who had solid employment. And then, without any prompting, Philemon declared, "Wilson is a magic man. He blows his horns to invite the whales." My jaw dropped. When I asked him about Eric Davalala, Philemon repeated, "Wilson is a magic man." He added, dismissively, "Eric just blows."

In contrast with Hermanus, Zwelihle has almost no tourists. But there were lots of people around, including groups of young men and women gathered in conversation outside storefronts and on street corners. Philemon told me that many unemployed young men in the township are addicted to *tik* (crystal meth); and, warning me to keep my distance, he explained that they have devised a peculiar form of theft. These young men blow a sleep-inducing powder into a house's window. Once the inhabitants are asleep, the thieves can then easily enter the house and help themselves. According to Philemon, the sleeping powder is made from crocodile bones. We seemed to have crossed, at some point, not only a geographical but an epistemological threshold.

Zwelihle is worlds apart from Hermanus in terms of wealth and appearance, though it lies a mere three miles to the southwest. Wilson's is a typical township home; row upon row of these small, square, "matchbox" houses were built during apartheid as dormitory cities for Black workers on the outskirts of cities and towns.[16] When Philemon and I arrived, two small children were watching TV in the living room, and one of them went into the bedroom to call his grandfather.

Wilson Salukazana the whale crier entered. He was a bit disoriented from having just woken up and was understandably surprised by my unscheduled visit. I explained to Mr. Salukazana that I was interested in his former work as a whale crier and that I would like to know more about it.[17] If Eric and Philemon had built up my expectations about Wilson being, or perhaps posing as, a magical communer with whales, these expectations were quickly dashed. Wilson recounted his past whale crying duties plainly: "You know when the whales come close by the shore, then I have to blow the horn, you see? . . . So that the tourists can come and take pictures with the whales. And some people forget about whale watching—they get busy with me. Because I had that regalia, uniform, with that hat, with that horn."

I would have looked to Philemon for support, but he was outside smoking a cigarette. I was weary of projecting fictional transformations of Wilson onto the man himself (now sitting in front of me, in the flesh) but given both Eric's and Philemon's testimonies, I felt it was reasonable to ask. So I stammered, "Some people say you used to call the whales with the horn?" Slightly surprised by my question, he replied, "Yes, I mean, that is a technique—the technique that I used to use. I don't know how did it happen, but it *happens*. Because sometimes when I blow the horn, for that particular code, and then you see the whales springing up, jumping up, you know."[18]

The conversation that followed was awkward, to say the least. It would be dishonest to say that I went into the interview without an agenda. I hoped, of course, to hear about Wilson the "magic man" and not Wilson the tourist curiosity. But for the most part, at least if one discounts elliptical statements like the one above ("I don't know how did it happen, but it happens"), Wilson basically denied any direct communication with whales. This was in sharp contrast with Philemon, who insisted on Wilson's ability to communicate with them.

For example, when Philemon had returned from outside, and we were all together in the living room, I asked Wilson, "Do you think they know you? Do you think the whales know you from being there all the time?" To this, Wilson simply laughed, seeming uncertain how to respond, but Philemon was nodding vigorously, and then said, "Spiritually." He paused briefly before repeating, "Spiritually."

Later in the conversation, I asked Wilson if he misses the whales since retirement. He answered that he does, and he remarked that sometimes he drives down to the bay to watch them in the water. But this still did not tell me much, so I asked another question: "Do you have an emotional connection with the whales?" To this, Wilson responded flatly, "I don't have a connection with them." But then he said, "You know, it's just because—when you love a thing at heart, it always happens. It always happens, people then sometimes think that you've got a magic that you are doing. Because I used to [i.e., when he worked near them] love these animals at heart." He continued in a somewhat convoluted and confusing way: "You know, I was seventy-one years old when I got to New Harbour for whale watching. And I worked there until I was seventy-four years old. So well, maybe, I'm tired, that's why. . . . Well, I used to love them. I don't know what to say. I used to love these animals at heart. I still love them."

To my ears, this was anything but clarifying. How to square the declarations of love with the statement, "I don't have a connection with them"? Was the implication simply that the love only traveled in one direction, that it was unrequited? Or did he mean something else? Philemon, who insisted on Wilson's spiritual connection with whales, had a different explanation. Later that day when we were alone, he told me that Wilson is a Born Again Christian and therefore finds it necessary to distance himself from certain Xhosa traditions.[19] Philemon seemed to suggest that Wilson simply could not admit (to me or to himself, I was never certain) that he has a spiritual connection with whales. But for Philemon, the connection is obvious. As he told me, "In our tradition, we believe that in the water—we've got our ancestors there. That's why he blows the horn, so that the whale can hear the sound."

But Wilson never said anything like that, at least not directly. He did, however, allude to some kind of spiritual connection at the end of our conversation. After saying, "I used to love these animals at heart," he briefly paused and told a story. Later, on our way back to town, Philemon suggested that Wilson's story was intended as a kind of parable about ancestors told by a devout Christian who could not, in good conscience, allow himself to openly expound his belief in occult forces, especially not in public and on record. I will never know for certain whether Wilson meant the story this way. In any event, it concluded our meeting that day. I present the tale here in full.

> One day we were there by De Kelders. And it was a new skipper at the boat. I was sitting there with people from Italy, guiding them in the boat. . . . So there near De Kelders we saw the whales. But they were very close to the shore. Because you see when we watch we go to Kleinmond, Hawston, Sonesta Beach. Or we go to near De Kelders, therrrrre [he stretched out the word like a sung note], the place called Die Plaat. That's where you find them. So what happens is that the skipper was new. He didn't know how to judge the wave. So this man, what he did—we went right by the shore there, [and] the seas come suddenly behind us. . . . Big wave, bigger than this house. That day, I thought I was going to die. And then the wave hit the boat on the side, and as I was standing like this [Wilson demonstrates], then I fall right into the water. Luckily, he thought he must switch the boat off because of the propeller. Dangerous, heh? Because I swim right, right round—the water brings me back to the propeller.
>
> [Philemon exclaimed, "*Hayibo!*" (Unbelievable!)]
>
> So when I come there, lucky I didn't breathe. I was brave enough, you know. That's why we say we must believe to our ancestors. Maybe our ancestors were looking at me. Holding me.
>
> ["They were watching you. You were protected," Philemon interjected.]
>
> So I didn't breathe. After five minutes I came up behind the boat. When I came up, I peer, emerge—I see, "Oh, there is the boat, the handle." Then I push myself . . . And then I come up and they hold me and they pull me into the boat.

The story ended. Glancing at me, Wilson added, "*Ja,* I'm trying to say, it's like that."

It's like what, exactly? Straining our ears, let us try to make sense of the story. Consider the moment when Wilson mentions his ancestors: "I didn't breathe. I was brave enough, you know. That's why we say we must believe to our ancestors. Maybe our ancestors were looking at me. Holding me." Perhaps I am making too much of these brief, elliptical sentences. But is not the ability to hold breath under water—expertly, even virtuosically—precisely what whales do for their daily survival?

Research shows that during drowning or near-drowning, there "is frequently an attempt to surface, and gasping may occur. During this period large amounts of water may be swallowed."[20] Involuntary responses to immersion in cold water do "not seem to serve any beneficial purpose. Instead, [they] can represent the first, and possibly the greatest, threat to the lives of individuals immersed in cold water."[21] M. J. Tipton, among others, has observed that "sudden cold-water immersion produces an 'inspiratory gasp' in subjects, followed by uncontrollable hyperventilation."[22] Involuntary gasping occurs immediately following immersion in cold water and is largely responsible for drowning fatalities.

It is possible that Wilson Salukazana survived that day only because he did not gasp for air involuntarily after being flung into the ocean. In this sense, it matters very much that the ancestors who were protecting him are connected with whales. Of course, on the biological level his body could only remain what it had always been: that of a human man. Obviously, he could not have held his breath for two hours, as whales do. But on the day that he thought he was going to die, he held his breath rather than gasping, like his ancestors who were protecting him and holding him. It is this same breath that resonates in his kelp horn each time he raises it to his lips. Breath, in other words, both separates humans and whales and connects them.

As I was writing this chapter, I stumbled on the master's thesis of Dorothy Winifred Steele, who had accompanied Mda to Hermanus in 2001, just as he was conceiving the ideas that would become *The Whale Caller*. Steele's account of her day with Mda is uncanny in light of my own work.

> We left the town and took a drive up the coastline, expecting to encounter the local whale crier at every turn, but there was no sign. We stopped to ask passers-by, but they, too, had not seen William Salukazi [*sic*] that day. Disappointed, we made one more stop, this time at a petrol station. The attendants suggested we go to the Information Bureau, and gave us directions. We found to our dismay that, as it was Wednesday, William Salukazi would not be on duty. Apparently, as he "worked" over the weekends, Wednesday was his "day off." We were startled to be told that the whale crier called people, not whales to himself. In that case, Mda said, it was just as well that we had not met him because he did not need a crier.[23]

Reading this, I realized that I had unwittingly retraced Mda's own steps some fourteen years later. And I realized, too, how different my own subsequent path had been from Mda's. For I had met Wilson Salukazana (fig. 2.1), and he had told me that he loved the whales at heart, and, indeed, that he still loves them. And he had told me his story (was it a parable?) of falling off the boat and being saved. And I had met Philemon, who had told me more about this "magic man," perhaps even more than the magic man himself would dare to say.

FIG. 2.1 Wilson Salukazana. Photo provided by the author. Reproduced courtesy of Wilson Salukazana.

When Mda finally published *The Whale Caller*, he made a point of thanking the whale crier in his acknowledgments: "If Wilson Salukazana (the real-life whale crier of Hermanus) really did call the whales to himself, . . . there would be no story to tell." In other words, Mda expresses his gratitude to Wilson Salukazana for *not* being a true Whale Caller, because if a Whale Caller already existed then Mda would not have been able to invent him. But *contra* Mda, and considering Salukazana's complex persona, many rich stories remain to be told. That one small, hypothetical *if* harbors a series of profound riddles, a host of secrets. Kelp horns resonate across time and place, in that elusive gap between whale callers and criers, both real and imagined.

How much of what Wilson and Philemon told me involves Xhosa tradition, or belief? In fact, their comments are to some extent supported by the anthropological record, which notes a connection between ancestors and water creatures in Xhosa cosmology. For many Xhosa speakers, the ocean is a space of tremendous spiritual energy and power. In an article on water in African cosmology, the anthropologist O. B. Lawuyi quotes a "witch in the Xhosa village of Qunu" in the Eastern Cape. "After I had undergone training under the sea," the witch says, "I started immediately to acquire extraordinary powers. I started to apply what I have learnt from there."[24] More generally, large bodies of water—oceans, lakes, and rivers—are important in various Xhosa rites of passage, including circumcision rituals and initiation into the role of a diviner. The process of becoming a diviner is sometimes referred to as *ukuthwetyulwa*—"to be called under the river."[25]

In some cases, apparently, this calling takes the form of a song from ancestors in the water. Joan Broster provides the English translation of one such song beckoning a candidate diviner:

Oh hurry to the river!
Oh hurry to the river!
What is detaining you?
I am complaining!
What is detaining you?
Oh hurry to the river![26]

Exactly what kind of being calls the diviner from under a river's surface is more difficult to determine. According to some culture-bearers, when diviners are under the river, they encounter a snake-like creature whose role it is to "communicate the will of the ancestors to their descendants."[27]Although it is often referred to as a snake-like creature, the spiritual being known as *Nomlambo* or *Mamlambo* (from the word *umlambo* [river]), is not reducible to the reptile, the snake (*inyoka*). *Nomlambo,* according to one diviner, can transform "into many people, girls, young and old men, and so on,"[28] which means that we are dealing not with mere reptiles but with water-dwelling shapeshifters.

In his memoir, Mda himself speaks about a water-dwelling spirit, referring to Mamlambo as a "water goddess who lives in the Mzintlava River [in the Eastern Cape province of South Africa] but travels in lightning to visit other rivers, including the river that runs in a narrow valley between our own Dyarhom Mountain and the Siqikini Mountain."[29] He goes on to say that the "true Mamlambo is a beautiful goddess with the torso of a horse, the neck of a snake and the lower body of a fish."[30] A strain of Xhosa mythology considers Mamlambo to be a "highly dangerous" water spirit that takes the form of "whatever one desires most deeply"—for some, this may be "the shape of a beautiful woman."[31] Mamlambo, at least in one iteration, is like a Siren of Greek mythology.

Zakes Mda's novel, *The Whale Caller* makes no mention of any of this: nothing about Xhosa mythology, nothing about ancestors or mermaids. The protagonist of the novel is a vague and seemingly cultureless and originless person—a kind of deracinated vagabond. In our conversations, Wilson Salukazana spoke about whales in a different way, and he mentioned ancestors only elliptically. But if questions surrounding ancestors and mythology remain oblique, love is declared loudly. That is, both Mda and Salukazana point variously to love and desire as foci of the interspecies encounter. A question arises: does the love expressed by people like Mda and Salukazana actually have anything to do with whales? There is certainly some truth to the charge that *The Whale Caller* is not about whales at all, that the book fictionalizes whales to such an extent that they become mere stand-ins for human desire. (Recall the water spirit that takes the form of whatever one desires most deeply.) Similarly, it's unclear what Salukazana means when he says that he "loves the whales at heart." As far as anything like communication goes, he observes only that whales sometimes spring up when he blows a particular "code" on his horn. But he does not claim to understand why or how that happens. (Eric Davalala would say that it's an illusion, and that the breaching of the whales does not depend on the horn calls.)

A critic may argue that when Mda (as a novelist) or Salukazana (as a horn player or mere observer) talk about loving whales, the love is so one-sided, so much a projection of human, male desire, that the presence or existence of whales hardly matters. In this sense, the critic may complain, the scenarios are less about interspecies communication than unidirectional desire. And yet, as I emphasize throughout this book, the shift from "pure," semantic communication toward more elliptical concepts is a common and perhaps even inevitable tendency in the communicative endeavor. Music, love, and spirituality are all supplements in Derrida's sense of the word: they seem to be mere additions to communication, but they are actually essential to many forms of communication.

In the sense that it offers a different take on human/animal relations, this chapter acts as a counterpart to the previous one. Especially when it comes to matters of the environment or nature, writers in the global North often hope to find "non-Western" conceptualizations. And they do sometimes find them: think, for example, of the understanding of water in Xhosa cosmology described in this chapter, or of anthropological studies of the ways indigenous peoples in Alaska interact with whales. Such work may be illuminating in some respects, but any statement to the effect that "Xhosa people believe that . . ." or "the Inupiat see whales as . . ." runs the risk of extreme simplification or, worse, exoticism. Yes, Salukazana and Mda both come from Xhosa-speaking households, but to say that they somehow hold a "Xhosa belief system" is grossly inadequate. What would that even mean for the cosmopolitan Mda, who grew up with pan-African ideology, plays classical flute but loves jazz above all other kinds of music, received Latin tutoring from South African Communist Party leader Chris Hani, devotes much of his time to beekeeping and farming activism, and has lived on and off in Ohio for the past several decades?[32] Or what would it mean to Salukazana, who lives in a staunchly Christian and socially marginalized township on the outskirts of an eco-tourism hub? Most of the people in this book (myself included) inhabit multiple, overlapping worlds. Most of the people in this book draw on a range of capacities (historical knowledge, mythical thinking, imagination, aesthetic expression, and philosophical speculation) in their encounters with non-human beings. In writing about those people, I have been similarly inspired.

When it comes to human/animal relations, "belief" seems often to be equivocal and hard to pin down. Especially because we all relate to animals from such complex subject positions, identifying what we believe or think about animals is no easy matter. Here is the important part: we may struggle to find the right language to describe our relationships with animals, but many people have no problem saying that they love animals. Indeed, love was the theme that seemed to turn up time and time again as I was thinking about Mda's *Whale Caller* and as I was speaking with Salukazana. Love is one of humanity's most powerful and ubiquitous emotions, but it is also one of the least understood. Is this because love is, by definition, beyond understanding? Perhaps. The writer bell hooks once argued that our unwillingness to provide a rigorous definition of love is part of the reason we do not know how to love properly.[33] In any event, the preceding discussion suggests that human love for animals falls into different types. The Whale Caller in Mda's novel has a primarily erotic relationship with the whale, Sharisha, while Salukazana's relationship with whales is one of quiet admiration ("I love them at heart") and, perhaps, of corporeal and spiritual protection.

There are many other places to look.

Animal Erotics

I turn first to none other than John Lilly, to an aspect of his research broached but not elaborated in the previous chapter. In the mid-1960s, Lilly opened a research facility (the Communication Research Institute, hereafter CRI) on St. Thomas, one of the US Virgin Islands in the Caribbean. His plan to learn about dolphin intelligence by spending time with the creatures in as natural an environment as possible was bolstered when a young woman named Margaret Howe visited the facility and offered her services. As part of his injunction that understanding the mind of a dolphin requires learning to "live wetly," Lilly created an amphibious environment in his research facility where Howe could spend most of her hours in the water with three dolphins. Raised only slightly above the water's surface were a bed and a telephone. The bed allowed Howe to be in constant, tactile contact with the dolphins, stroking them as she reclined.

Howe lived with the dolphins for approximately ten weeks between 1964 and 1965, an experience that she documented meticulously and that caused something of a scandal at the time.[34] The scandal has since attracted scholarly attention and was the subject of a BBC documentary (*The Girl Who Talked to Dolphins*), in which Howe was interviewed some forty years after the fact. The details of the episode are embarrassing for those involved, and most historians have largely avoided them. (When I presented this material at Princeton University in 2017, Scott McVay—co-author, with Roger Payne, of the seminal article "Songs of Humpback Whales"—was in attendance, and he objected strongly to my speaking publicly or writing about the episode. If the episode is embarrassing, however, that is because of the actions of those involved and not because of my writing about it.)

Of the three dolphins at the CRI, a young male named Peter featured most prominently in Howe's notes. Reading Howe's daily entries and listening to the recordings of her interactions with the young dolphin reveals two gradually diverging paths. First, we note Howe's growing frustration with Peter as the poor dolphin fails, time and time again, to speak intelligibly. Lilly had always maintained that the only way to establish human/dolphin communication was to *believe* that it is possible: "We have found that, in dealing with such a large-brained mammal, we must keep the working hypothesis in mind that 'they are highly intelligent and are just as interested in communicating with us as we are with them.' If we use any other hypothesis, we have no success whatsoever in dealing communicatively with them."[35] He insisted that Howe act like a mother—presumably with all the grit and unconditional love he expected of one—if she wanted to teach young dolphins to speak English.[36]

Howe's experience with the dolphins shows that she clings to Lilly's words, even as the possibility of success slips further from view. A typical entry reads, "I am well pleased with all of Peter's activities except his apparent vocal ability. He is more than eager . . . works hard . . . but he just does not seem to hear or be able to copy the pronunciation aspects of speech. Perhaps this will come . . . perhaps not." Later: "Several of the lessons this week were very poor. Peter has picked up this monotonous tone, a whine. . . ." Later still: "I still feel very strongly that I can do what I am trying to do, . . . but I have not succeeded in doing it yet. I must teach Peter that he is to learn."

And then there is the second, diverging path. Howe's confidence in Peter's vocal ability wavers, but she begins to develop a deep affection for him. Nothing can substitute for her own words:

> Peter is his energetic self and a bit nippy on the toes. I carry a long-handled broom with me for that and ward him off. This is not always the case, of course; we have had several long "loving" sessions. The water is deep enough for him to roll over and this he does for tummy rubs. (June 15–19, 1965)
>
> Several new things developed this week. First, Peter and I got on a more sociable, physical level. Peter began to be gentle with me, and allow me go to him without the broom. . . . Doing this he is very gentle . . . the rough part being when he tries to push me around. He gets between my legs and pushes me apart. (June 19–25, 1965)
>
> Peter has become sexually aroused several times during the week, and I have thoughts and questions on this that follow in a separate paper. (July 3–10, 1965)
>
> Peter is courting me . . . or something very similar! I begin to take an active part in the play. . . . [. . .] I feel extremely flattered at Peter's patience with me in all this . . . and am delighted to be so obviously "wooed" by this dolphin. (July 17–August 1, 1965)
>
> Now it has happened that Peter has modified his sexual rambunctiousness . . . to a more humanized level . . . and no longer has me come to a dead stop when he gets excited. [. . .] I can very easily rub his penis with either my hand or my foot. Peter accepts either and again seems to reach

> some sort of orgasm and relaxes. We usually go through this three times or so before he quits and starts another sort of game. This is not a private thing. Peter and I have done this with other people present . . . but it is a very precious sort of thing, Peter is completely involved, and I involve myself to the extent of putting as much love into the tone, touch, and mood as possible. (August 2–18, 1965)[37]

Howe goes on to explain that Peter taught her about sexual trust and that he alone brought her out of a state of depression on several occasions. In *The Girl Who Talked to Dolphins,* Howe recalls that she and Peter "just hit it off" and that the dolphin was "madly in love" with her. Howe ultimately married the photographer hired to take photos of her and Peter at St. Thomas. About Peter, she laments, "We couldn't elope, we couldn't rush off into the sea and hide."[38]

Peter died shortly after the separation from Howe, and Lilly told Howe that Peter had committed suicide. Hyperbolic as it may sound, bereft dolphins are known to deliberately kill themselves. How? A dolphin can only sleep with half of its brain at a time (a process known as unihemispheric sleep), the other half being required to take care of functions such as simple swimming and replenishment of oxygen from the air. This means that dolphin breathing is to some extent always voluntary, and experts believe that dolphins commit suicide by deliberately ceasing to breathe. If Peter did kill himself, though, it was not necessarily because of his separation from Howe, but more likely due to other factors: at the conclusion of Howe's experiment with Peter, the CRI closed down, and Lilly moved Peter to a new facility in Florida. In contrast to the relatively spacious CRI, Peter was kept in a small tank with little sunlight.[39] Attributing Peter's suicide (or at the least, his giving up the will to live) to his yearning for Howe is dubious at best. Interpreting Peter's suicide as a response to being separated from Howe probably reflects Lilly's and Howe's narcissism more than anything else.

For the purposes of this book, the question is this: What kind of love was the sexual encounter between Margaret Howe and Peter? And in what sense was that encounter a kind of communication, if it was one at all? To begin answering the question, let's consider Lilly's own response. Lilly had long maintained that "making love" constituted a human "form of exchange and communication."[40] This idea was severely hobbled by his homophobia, however. He believed that mental health was characterized by, as he put it, "fulfilling sexual experiences in the *heterosexual sphere.*"[41] As he wrote, "One such area is a common human communication path. That is the part of the man and woman in love, making love."[42]

If only sex between a man and a woman counts as "communication and exchange" in Lilly's books, then what of sex between a human woman and a male dolphin? Posing the question of sexual communication between humans and animals, Lilly said only that "on the human side this is called a crime (sodomy)."[43] Indeed, Western legal systems have often applied the term *sodomy* to various forms of outlawed sexual acts, including homosexuality and bestiality. Because he does not dwell on the topic beyond the fact of its being a crime, Lilly's view of the "human side" of human-animal sexual relations remains vague.

To his own question whether "sexual activities [might] be used for communication across the interspecies barrier," he replied, "On the delphinic side, the answer would probably be yes."[44] Lilly implicitly condones Howe's treatment of Peter throughout *The Mind of the Dolphin*, and it seems that Lilly even saw potential scientific benefit in sexual relations across species lines. The issue of delphinic consent seems never to have occurred to the scientist, and it behooves us to pose the question now, at least if we are to develop any kind of ethical position on the matter.

One of the reasons that Lilly ignores the question of consent is that Peter was a male dolphin who, at least superficially, seems to have asserted himself on Howe (sometimes against her will). Other writers on the topic, including the famous philosopher Peter Singer, seem to deploy the same logic: If a male animal seems to desire sexual contact with a human (especially, it would seem, a human woman), then the question of consent from the animal is deemed irrelevant. Unsurprisingly, Singer's argument is more complex than Lilly's, as we will soon see. But both make the mistake of assuming that apparent male arousal is equivalent to consent.

The philosopher Amia Srinivasan observes that, "the conflation of pleasure and consent is commonplace in rape apologism. It's at work whenever an orgasm or an erection is taken as proof that someone consented, even when they insist that they didn't—or, as with young children, when they are incapable of meaningful consent."[45] She quotes a 2005 article by Gieri Bolliger and Antoine Goetschel, who observe that, although exceptions may exist, animals who have sex with humans are typically trained to do so.[46] Srinivasan continues, "Even self-identified zoophiles often 'groom' animals: bribing them with food, constructing specialised barns, halter breaking, acclimatising them to human penetration with dildos, injecting them with hormones. Such conditioning, Bolliger and Goetschel write, not only infringes 'the free sexual development of an animal,' but runs the risk of creating a 'strong dependency' of the animal on the human. Thus the 'violation of an animal's sexual integrity . . . does not depend upon the question of what an animal feels during a zoophilic act.'"[47]

To speak of consent in the case of a captive dolphin like Peter makes no sense at all. Indeed, sexual violence was committed against him, and not the other way around. To put the whole thing in broader context, consider the construction of the CRI where Peter was caged. Lilly seemed to view himself as something of a modern conquistador, and in choosing the location for the Saint Thomas facility, he wrote, "I heard that this location, once cleared, must be a magnificent one because the old Danes had chosen the best locations when the island was cleared for the growing of sugar cane. They had the advantage over us of immediately available slave labor to clear vast acreage quickly and thus selected their building sites in the clear."[48] Here, Lilly literally bemoans the fact that he did not have the "advantage" of slave labor. And yet, as he noted, founding the institute was only possible because of money from his wealthy second wife, Elisabeth Bjerg, whose "grandfather was the royal Danish customs collector at Frederiksted on the island of St. Croix before the islands were bought by the Americans."[49] While he did not have actual slave labor at his disposal, Lilly did benefit from money from a colonial family (and likely from a fortune amassed during slavery). What is more, despite his reliance on labor by Black Caribbeans at the CRI (a fact confirmed by photographs), Lilly never once—at least not that I have seen in his dozen or so books, and in thousands of pages of his personal papers currently housed at Stanford University—mentions a single Black laborer, even in his detailed description of the institute and its personnel. Given everything I've reported about Lilly, this must come as no surprise. It was in that context that Lilly held dolphins captive, and it was in that context that he hired a young woman (a waitress, *not* a scientist) to "communicate" with Peter. To describe the relationship as one of "love" is an abuse of the term.

Peter Singer's brief paper on human/animal love begins from a different perspective. As a longtime advocate of animal rights (his 1975 book, *Animal Liberation*, basically sparked the movement), Singer argues against the moralizing impulse to denounce human/animal sexual relations as something that would offend "our dignity as humans."[50] He writes, "The taboo on sex with animals may have originated as part of a broader rejection of non-reproductive sex. But the vehemence with which this prohibition continues to be held, its persistence while other non-reproductive sexual acts have become acceptable, suggests that there is another powerful force at work: our desire to differentiate ourselves, erotically and in every other way, from animals."

This motivation is, for Singer, not sustainable either on philosophical or ethical grounds. Furthermore, the argument that bestiality is shunned because it is cruel seems unconvincing to the lifelong advocate of animal rights and veganism. Yes, Singer admits, sex with animals is often cruel (the examples he provides are grotesque and upsetting), but so is pumping caged chickens full of hormones and then cutting off their beaks and heads, as the poultry industry has been doing for decades. He argues, therefore, that the sexual taboo cannot be based purely on a prohibition against violence; it must have a different foundation.

What, then, of cases of human/animal sexual relation where cruelty seems not to be at issue? What of cases when both parties seem to consent? In such cases, Singer suggests, surely the reason for a taboo can only be moral: an offense to our dignity as humans. But the question remains whether a consensual sexual relation between humans and animals can exist, and here we need to briefly examine Singer's two examples.

His first example is of an orangutang in Borneo that, apparently, physically seized a human woman, "his intentions made obvious by his erect penis." The incident happened at Camp Leakey, "a rehabilitation for captured orangutangs," and while Singer does not provide details about the exact status of the particular orangutang, we know from his story that the ape was either in captivity at the time or was in the process of rehabilitation after being held in captivity. In other words, the woman was not simply meandering through the forest when the orangutan seized and threatened her.

Singer's second example is a familiar one. "Sex with animals does not have to be cruel," he writes, and then continues: "Who has not been at a party disrupted by the household dog gripping the legs of a visitor and vigorously rubbing its penis against them? The host usually discourages such activities, but in private not everyone objects to being used by [a] dog in this way, and occasionally mutually satisfying activities may develop." Does this example prove his point about consent?

Singer describes what animal behaviorists term *mounting* and *humping*. There is no consensus as to why dogs hump people (or other things), and there seem to be several motivations for the act. Male and female, intact or sterilized, dogs hump. They do so, according to experts, when they are aroused (sexually or otherwise) as well as when they are anxious. Sometimes dogs hump to establish dominance (although this motivation seems to be less frequent than pet owners often assume). Humping can also be attention-seeking behavior or a form of play.[51] Experts often emphasize that humping is not necessarily, or even most commonly, sexual.

Veterinary scientists and dog trainers recommend against generalized theories, but play seems to offer a particular point of illumination, at least if we understand play in a somewhat technical sense and not simply as akin to having fun or being silly. "When, for example, dogs play together," writes anthropologist Eduardo Kohn, "they act as if they are fighting. They bite each other but in ways that are not painful." Kohn goes on to quote Gregory Bateson: "The playful nip denotes the bite, but it does not denote what would be denoted by the bite." It is as though the dogs are saying, "These actions in which we now engage do not denote what those actions *for which they stand* would denote."[52] We can paraphrase Bateson and Kohn like this: humping denotes sex but it does not denote what would be denoted by sex. Or, to state things another way, a playful nip is to a bite what humping is to sex. In one respect, then, the many scientific and journalistic articles warning pet owners against simple explanations regarding *why* their dogs hump miss something important. For, irrespective of whether a dog humps because it is anxious, aroused, or seeking attention, the act of humping denotes sex (it is mimetic of sex)—nothing is gained by denying this.

In another respect, because humping merely denotes sex but is not itself sex, its motivations may be different from what motivates sex. If indeed nipping/biting is structurally analogous with humping/sex, then the questions, "Why do dogs pretend to bite when they have no intention of biting?" and "Why do dogs pretend to have sex when they have no intention of having sex?" should be replaced by, "Why do dogs play?" In other words, why do dogs engage in behaviors that denote, at one step removed, instinctual behaviors? This is an exceedingly difficult question to answer, but even posing it sheds light on the topic. Play points, or begins to point, at a complex semiotic endeavor in which meaning is at once obfuscated and communicated—or, better yet, communicated dimly, as a kind of shadow. Play marks the beginning of a rich form of semiosis where a gap between intention and action is opened. It is a nascent play of the imagination, if not quite or not yet a free play of the imagination in the sense intended by Kant.

In any event, humping is a poor example to choose when trying to prove that animals sometimes want to have sex with humans. Despite the lack of consensus about humping, the behavior is so variegated and so complex that few, if any, researchers view it as a genuine attempt at a sexual encounter. (It seems in most cases, as I have suggested, closer to play.)

In relation to the other examples (of Peter and of the orangutan at Camp Leakey), a question remains: Are dogs captive? If we understand the dog to be a domesticated animal (this is commonly part of the definition), then in some sense dogs are indeed captive: whether they are kept indoors (increasingly common in many parts of the world), within an enclosed yard, or in a fenced-off farm, dogs are almost by definition animals whose mobility is circumscribed by humans. Dogs are not captive in the sense that a dolphin in SeaWorld (or a scientific lab) or an orangutan in the zoo (or a scientific lab) is, however. This is because dogs have been domesticated over many thousands of years and have become what Donna Haraway calls "co-species companions."

Dogs were probably the first domesticated animals; through the process of domestication, they developed "distinct morphological and behavior traits."[53] In particular, dogs exhibit high levels of social tolerance and attentiveness toward humans in ways that wolves, for example, do not.[54] These behavioral traits evolved over thousands of years—a process of natural and "artificial" selection under pressure from various human communities, including breeders.[55] Despite the significant morphological and behavioral changes that this process has wrought, (domesticated) dogs and (wild) wolves are "nearly identical at the level of gene sequence,"[56] and can frequently produce healthy offspring. Genetic differences between dogs and wolves seem rather to happen at the level of gene expression (that is, epigenetic modification of gene activity) and in the so-called "noncoding" parts of DNA.[57] This is a thorny and relatively new domain of inquiry, which is remains only partially understood.[58]

One recent study out of Iran found that several factors are frequently associated with humping and mounting. Dogs are more likely to exhibit unwanted humping behavior when: (1) they have been owned for longer than six months, (2) they are kept indoors, and (3) they live in a household without other dogs. It is not difficult to imagine why a dog kept indoors and away from other dogs for a long period of time would exhibit unwanted behavior.[59] This study and others like it imply that the behavior is more likely motivated by anxiety borne from being kept alone inside and in "captivity"—a response at odds with the one Singer suggests, which describes humping as an attempt to establish a "mutually satisfying activity" (his euphemism for something overtly sexual). In other words, humping may express a desire for "love," but probably a kind of love other than the one Singer imagines. It may explain how humping, as a playful behavior denoting sex, is motivated by anxiety and loneliness.

When all is said and done, eroticism may not be the best place to begin when trying to understand the human/animal relations. What other kinds of human/animal love may be imagined? Recall that when Lilly invited Howe onto the island, he expected her to treat Peter in the way a mother would treat a son. The mother-son relationship was meant to foster language acquisition, but of course something else entirely happened.

In their magnum opus, *A Thousand Plateaus*, philosophers Gilles Deleuze and Félix Guattari argue that there are three kinds of animals, and they reserve their most scathing words for one kind: "individuated animals, family pets, sentimental, Oedipal animals each with its own petty history, 'my' cat, 'my' dog. These animals invite us to regress, draw us into a narcissistic contemplation, and they are the only kind of animals psychoanalysis understands, the better to discover a daddy, a mommy, a little brother behind them (when psychoanalysis talks about animals, animals learn to laugh): *anyone who likes cats or dogs is a fool.*"[60] It is true that in many cases house pets stand in for familial relationships, but elsewhere Deleuze admits that it is possible to have a different kind of relationship with one's cat or dog. In *Deleuze from A to Z*, he begins with "animal" ("A is for animal"), noting that while some people have *human* relationships with their pets, it is also possible to have an *animal relationship with an animal*, including a cat or a dog.[61]

Donna Haraway similarly notes that, in some contexts, dogs are used as "alibis for other themes."[62] "Commonly in the US," she writes, "dogs are attributed with the capacity for 'unconditional love.' According to this belief, people, burdened with misrecognition, contradiction, and complexity in their relations with other humans, find solace in unconditional love from their dogs."[63] On this point, Haraway is every bit as uncompromising as Deleuze and Guattari: she calls the notion of a dog's unconditional love "abusive" and "pernicious," arguing that it is "based on a lie."[64] Instead, Haraway uses the example of Linda Weisser, breeder of Great Pyrenees livestock guardian dogs, as a model. Haraway quotes some potent lines from Weisser, a few of which I reproduce here.

> I love this bitch beyond words. [The root for my love is about] the deep pleasure, even joy, of sharing life with a different being, one whose thoughts, feelings, reactions, and probably survival needs are different from ours. . . . While my dogs can love me (I think), I have never had an interesting political conversation with any of them. On the other hand, while my children can talk, they lack the true "animal" sense that that allows me to touch, however briefly, the "being" of another species so different from my own with all the awe-inspiring reality that brings me.[65]

To invoke Deleuze's terms, Weisser establishes an animal relationship with her dog rather than a human one. Taking a cue from Weisser, Haraway writes, "I resist being called the 'mom' to my dogs because I fear infantilization of the adult canines and misidentification of the important fact that I wanted dogs, not babies. My multi-species family is not about surrogacy and substitutes; we are trying to live other tropes, other metaplasms. We need other nouns and pronouns for the kin genres of companion species, just as we did (and still do) for the spectrum of genders."[66] Indeed, even the word "family" in the above passage is not quite right, as Haraway herself would be the first to admit. The maxim of her entire philosophy—"make kin, not babies"—is not about treating non-baby kin (dogs, for example) as though they were one's babies.[67] On the contrary, it is about finding ways of relating to dogs as dogs.

Despite their varying standpoints, Weisser, Haraway, and Deleuze and Guattari all allow for the possibility of love. "I love this bitch beyond words," writes Weisser, even though she admits to the impossibility of conversation, or even smooth communication, between herself and her dog. Deleuze and Guattari mostly hate pets, so their example of human/animal love is far darker and more fantastical: they write about a character named Willard (the protagonist of a film of the same name) who breaks all human ties and makes the "sinister choice" to form "a kind of contract or alliance, a hideous pact" with a rat.[68] The sequel to *Willard* ends with a young boy named Danny being reunited with his pet rat, Ben. As he cradles and strokes his rodent companion, the voice of a young Michael Jackson croons on the soundtrack: "Ben, the two of us need look no more / We both found what we were looking for / With a friend to call my own / I'll never be alone / And you my friend will see / You've got a friend in me." This is a deliberately bizarre example, one where the innocent voice of Michael Jackson as a child works uncomfortably against the grain of the boy Danny caressing his rat. The scene takes on additional unsettling overtones when we view and listen to it today, knowing everything we do (or think we do) about Michael Jackson as an adult. Many forms of love are present, all of them disquieting.

Haraway, too, loves her animals. She writes about her dog, "We are training each other in acts of communication we barely understand. We are, constitutively, companion species. We make each other up, in the flesh. Significantly other to each other, in specific difference, we signify in the flesh a nasty developmental infection called love. This love is an historical aberration and a naturalcultural legacy."[69] Love as a nasty developmental infection: to live together domestically means learning to love one another. This is not an erotic love, nor necessarily a sentimental one. It is something much simpler, but also maybe more challenging.

On this note, I take inspiration from the science-fiction author Ted Chiang, whose work sometimes explores the contours of love between humans and non-humans. One of his short stories focuses on the relationship between humans and AIs (artificially intelligent machines) that have acquired some measure of consciousness through human-led training. Some of what Chiang says in his author's note following the story applies to animals as well. In writing the story, he says,

> I was also interested in the idea of emotional relationships between humans and AIs. And I don't mean humans becoming infatuated with sex robots. Sex isn't what makes a relationship real, the willingness to expend effort maintaining it is. Some lovers break up with each other the first time they have a big argument. Some parents do as little for their children as they can get away with. Some pet owners ignore their pets whenever they become inconvenient. In all of these cases, the people are unwilling to make an effort. Having a real relationship, whether with a lover, or a child, or a pet, requires that you be willing to balance the other party's wants and needs with your own. I've read stories in which people argue that AIs deserve legal rights. But in focusing on the big philosophical question, there's a mundane reality that these stories gloss over. It's similar to the way movies always depict love in terms of grand romantic gestures, when over the long-term love also means working through money problems, and picking dirty laundry off the floor. So, while achieving legal rights for AIs would be a major step, another milestone, that would be just as important, is people putting real effort into their individual relationships with AIs.[70]

Even as legal cases continue to be built in support of "non-human rights" (or non-human rights to personhood), Chiang's words sound a fresh clarion call. As bell hooks would say, "Love is a verb."[71] And for Chiang, the verbal form of love is simply making an effort.

Alex & Us

Another story—the story of Alex the parrot—bifurcates in a manner that by now will be familiar. On the one side, there is the narrative displayed not only in a scientific monograph and articles, but also in various newspapers and television shows. Using a novel approach, researcher Irene Pepperberg managed to overturn a long-held view of parrots—namely, that they use human words in a purely imitative fashion. Going against the tide, Pepperberg convincingly demonstrated that one particular parrot (Alex) had learned to do much beyond imitation. Alex could, for example, identify colors, shapes, and numbers by sight. He also seemed to grasp more abstract notions such as zero.

The other side of the story is less tidy. Yes, Alex could identify (and state through words) concepts and categories, but his communication and his interactions with humans amounted to far more (and perhaps also far less) than that. In addition to her scientific work, Pepperberg wrote a best-selling memoir about that work, and it is there that we learn about her relationship with Alex beyond rudimentary pedagogy concerning numbers and shapes. The full title of the book, comprised of two subtitles and twenty-four words in total, is indeed telling: *Alex & Me: How a Scientist and a Parrot Discovered a Hidden World of Animal Intelligence—and Formed a Deep Bond in the Process.*

Because she worked with Alex for decades, Pepperberg developed a deep bond with the bird (to parrot the subtitle, as it were), even if this was not her intention. In fact, she tried hard *not* to let this bond develop and only recognized her deep feelings for Alex after his death. She writes, "The sense of loss, grief, and desertion that tore viscerally at my heart and soul at the passing of my one-pound colleague and companion of three decades was of a degree and intensity I had never anticipated, nor could have imagined. A huge torrent of love and caring, assiduously kept in check by a solid dam for all that time, suddenly burst through; the liberated flood of emotions swept all reason before it. I have never felt such pain nor shed more tears. And hope never to again."[72]

Alex's last words to Pepperberg? "You be good. I love you" (202). *Alex & Me* begins with the death of the celebrity parrot and recounts the outpouring of sorrow that followed, not only from Pepperberg, but from Alex-admirers far and wide. Because nothing can substitute for the original words, I present a selection below, as recounted in *Alex & Me*.

> "I cannot begin to tell you how deeply I feel for you," wrote Mother Dolores Hart. "I am a member of the Abbey of Regina Laudis community of Benedictine contemplative nuns, in Bethlehem, Connecticut. We have a Grey who we love so much and who has kept me in close awareness of you and your work. Your sudden loss is heartbreaking. We will keep you in our prayers and our love of these wondrous creatures who have shown us something more of God than we could have ever believed possible." Mother Hart is the abbey's prioress, a former Hollywood actress who had the distinction of playing opposite Elvis Presley in two films and who starred in the 1960 classic *Where the Boys Are*. She left the glamor of film and entered the Abbey of Regina Laudis four decades ago, and has had the company of an African Grey for the past seventeen years.

A woman from Michigan wrote to Pepperberg after hearing about Alex, "I just wanted to write this and say that I have been suffering from a clinical depression for weeks, very numb inside despite being surrounding by a loving family and nearly two hundred beloved pets, my children, ands grandbabies. Since I learned of Alex's death I have been reading the e-mails on the Web site, and tears are finally flowing. . . . I forgot how to feel emotions, and reading Alex's nighttime words to his dearly beloved friend [i.e., 'I love you'] opened the floodgates for me." An elementary school child from Illinois wrote, "I feel really sorry that your friend Alex left you. But he's in a better place now." And another wrote, "Alex must have meant a lot to you. He will always be with you. I lost my grandmother a few years ago. But deep down she is always with me. Just like Alex is with you" (20–21).

The book begins with Alex's death, and then moves back in time to Pepperberg's childhood in subsequent chapters. There we learn that she came from an unloving home with no siblings and with parents who were unable to speak about the past or, indeed, about anything of substance. In that context, the socially awkward Irene Pepperberg was thrust into a school where she "was the only white child in the class," and where she dreamed about finding a talking parrot like Polynesia from *Dr. Dolittle*. Needless to say, she found exactly such a parrot in Alex.

Pepperberg's account of her childhood shows that the desire for a talking parrot was intended to fill in a central lack. Because her parents would not talk to her, and because she was an outcast at school, she turned simultaneously to books and to birds. Her lifelong quest to teach a parrot to speak English is narrated as if by a mother, thus fulfilling Lilly's and Howe's dream, only with an avian rather than an aquatic companion.

At the end of the book, Pepperberg returns to Alex's death. "I took care of Alex, as any dutiful Grey owner would," she writes, "but he was such a free spirit that I never felt I *owned* him." Here, in the final pages of the book, she turns to a rather surprising analogy. The feeling of never owning what you love, she says, is "best expressed in one of my favorite films, *Out of Africa*." Based on a book of the same name, *Out of Africa* tells the story of a love affair between Danish baroness Karen Blixen and big game hunter Denys Finch-Hatton. It takes place in British-occupied East Africa. Pepperberg writes, "The book opens with the simple and yet deeply evocative phrase, 'I had a farm in Africa.' It's difficult to explain, but when you go to Africa, the place gets under your skin, burrows into your soul. And so that simplest of opening lines instantly taps into the most fundamental of emotions." This analogy is as disconcerting as it is surprising. For all her scientific rigor, the book ends with a base colonial cliché.[73]

In the concluding lines of the book, Pepperberg expresses her strong identification with the baroness on two fronts: first, because she was not properly recognized by the men of her day (this is a reference to Pepperberg's struggles to attain funding and secure employment); second, because the baroness mourns the loss of something she never truly owned—for Blixen, the hunter (played by Robert Redford in the film); for Pepperberg, her African Grey, Alex. Although Pepperberg draws an analogy between Alex and the hunter as two objects of desire and deep longing that could never be captured, it seems to be Alex's home—Africa—that emerges ultimately as the truly mysterious and unknowable entity of the book. Despite decades of teaching Alex to be a linguistic creature, in the end she could not *own* him—this is Pepperberg's message. And in meditating on her inability to own him, she observes that the continent his species was violently captured from is beyond words. It is difficult to explain what Africa feels like, she says. Africa gets under her skin and burrows into her soul. At the end of a book largely about words, words fail her.

Alex & Me pithily captures a core aspect of this chapter, namely, the forking paths of human/animal interaction: one path leads to the prosaic level of "communication" (Alex identifying the color red, for example), another to deep love and affection, where a motley crew of nuns and schoolchildren weigh in, and where only the analogy of a film about colonized Africa starring Robert Redford and Meryl Streep will do. Here again, we witness less-than-subtle sexual overtones. Comparing Alex not with just any actor but with Robert Redford certainly seems to make the subtext text. And even if it is Finch-Hatton and not Redford she is comparing Alex to, that book is one of the most famous doomed love stories of the last century. The analogies are unsubtle, perhaps even overwrought.

One other aspect from *Alex & Me* requires further comment. In thinking back about Alex's life, Pepperberg reflects for a moment on all the captive parrots of the world—those kept as pets, as well as those in scientific laboratories. In a conversation with Marc Johnson from the organization Foster Parrots, Pepperberg made the following analogy: "It is like putting a four-year-old kid in a playpen in the morning and leaving him there all day by himself." Johnson wholeheartedly agreed. "You have to remember that dogs have been domesticated for thousands of years," he told Pepperberg. "Mostly they are fine being in a home environment by themselves. But parrots are not domesticated. They are wild animals, and we shrink their world to the size of a single room in your house, or even to the size of a small cage. We have to remember that. That's why we need ways of entertaining them when you aren't there" (168–69).

How to solve the problem? It was during her time as a fellow at the

MIT Media Lab that her colleague, Ben Resner, came up with a novel idea. "Why not find a way for parrots to be able to surf the sites they find interesting?" he asked, apparently half in jest. Resner's provocative suggestion, along with Pepperberg's sustained meditation on the topic, led to the advent of the animal internet. The launch of the animal internet at MIT is the same event described in the introduction to this book, during which Jane Goodall talked about her dog Rusty, and Peter Gabriel spoke about playing music with apes. Reflecting on its early conceptualization with Resner at MIT, Pepperberg notes, "The idea ultimately came down to Alex, or whomever, being able to select between four modes using a joystick: pictures, music, games, and video. Within each mode were four choices. In music there were clips of a classical piece, a rock song, jazz, and country music" (169).

Pepperberg's work with Alex showed that parrots seem to exhibit preferences for certain kinds of music. A substantial amount of subsequent research has been done on this topic, most of it motivated by scientists looking for ways to keep their captive subjects occupied. One recent study claims to have found that chimpanzees prefer West African (Akan) music and North Indian classical music to silence. The paper is so riddled with problems that the reader can hardly take it seriously.[74] What does seem to be true is that captive chimps spent more time in a part of their enclosure where African or Indian music was being broadcast via speakers.

The image stays with me: captive chimps hewing to the sides of an enclosure where a recording of human music-making is most audible. Does this show a preference for certain melodic and rhythmic structures, as the authors claim? Or does it not instead signal something else—an attempt to alleviate loneliness and boredom by these magnificent creatures held captive in Georgia? (Recall those poor elephants in Paris, who were subjected to a similarly naive and cruel musical fate.)

But if chimpanzees and parrots are being held in cages by people who attempt to find out what music may help to relieve their desperation and loneliness, those same people are often attempting to escape their own loneliness. Those same humans may be attempting to escape other humans through relations with the animals they hold captive. Consider Margaret Howe out looking for an animal adventure before a photographer whisked her away, Irene Pepperberg overcoming her childhood trauma by creating an avian conversation partner, and the fictional Willard seeking the company of rats.

And then I think of other examples. I think of Timothy Treadwell, immortalized by Werner Herzog's film *Grizzly Man*, who overcame his substance abuse problems and depression by spending summers in Alaska with grizzly bears. Treadwell was deeply ambivalent about humans (he declares his misanthropic feelings throughout the film), but he declares his love for bears and foxes incessantly. "I love you," he says to the bears that cross his path. "I am so in love with all of you." Treadwell, famously, was eaten by a bear.

Or I think of Gavin Maxwell with his pet otters in Scotland—his love for these creatures forming a sharp contrast with the anxiety he experienced around humans.[75] Or I think of Craig Foster, whose documentary, *My Octopus Teacher*, brought the effervescent world of these creatures into millions of Covid-lockdown-era living rooms. Foster, famously, not only loved an octopus, but learned from her how to cope with the world of human anxiety and stress.

And I think again of Zakes Mda's creation, the Whale Caller. The Whale Caller, too, is someone who struggles with human relations and harbors misanthropic feelings. Especially when things are going poorly with his human girlfriend (an unstable alcoholic named Saluni), the Whale Caller heads to the cliffs, where he forlornly blows his horn. Does Sharisha the whale hear him? Does she truly respond to his love calls? Or is he deluding himself—like Timothy Treadwell? Commenting on the film *Grizzly Man*, Werner Herzog says he fundamentally disagreed with Treadwell's interpretation. Where Treadwell saw great beauty, looking into the eye of a grizzly bear Herzog sees only "a half-bored interest in food."[76]

And then there is Wilson Salukazana, whose views seem more balanced, and who, like most of us, puzzles at the world around him. He admits that the whales seem to respond to his horn calls, but he also admits to not knowing exactly why. He is willing, at least in one vulnerable moment, to entertain the possibility that his ancestors protected him when he nearly drowned. According to Philemon, those ancestors are certainly connected to whales.[77]

And Wilson Salukazana, in the midst of his equivocations about whales, can say (in a way many of us feel that we can): "I love them." When it comes to our relationship with animals, it would seem that love is the one thing a lot of people can agree upon. Love can signal misanthropy—a love of animals can be a salve for the hatred of humans. But it need not be. Sometimes love is what we say when we want to affirm a relationship without knowing how to describe it otherwise.

Consider this: a 2016 report on the relatively frequent practice of adoptive parenting among marine mammals ends in bewilderment. The authors of the report puzzle over the "costly" caregiving activities of Indo-Pacific dolphins and conclude that it is "unclear why an animal would invest its resources in this manner."[78] To this, Alexis Pauline Gumbs responds in her extraordinary book about marine mammals, "Who has not been mothered by someone genetically and socially distant from your birth situation, at some necessary time? And if you have ever shared something, taught someone, shared responsibility for someone's wellness for even a part of their journey, how would you measure what you gained from that potentially 'costly behavior'? We call it love."[79] Margaret Grebowicz reminds us that "the Internet now brings us daily updates about interspecies relations in the wild, especially in life-threatening situations (lioness saves baby deer from other lions, orangutan saves baby bird from drowning), but in others as well (polar bear pets dog, cat loves dolphin)."[80]

In Gumbs's writing, love does not explain anything so much as it invites us to move beyond scientistic reductionism. Love is not a form of measurement, but a resistance against mismeasurement, and perhaps even against measurement itself. "And think about it," Gumbs continues, "scientists, especially those people who have designed their entire lives around the hope, the possibility that they will encounter a marine mammal, and who have taken extreme measures (like moving to Antarctica) in order to increase the likelihood that they will see some particular being, cannot be unmoved. They are clearly obsessed, and most likely, like me, in love. Whether they admit it in their publication or not."[81]

"I love you," says Gumbs to the dolphin, the orca, the walrus, the marine otter, the sea lion, and the seal. And to the whale, so beloved for its musical ability: "I love the parts of you that no one thinks are particularly special. I love the basic you of you unmarketable and everyday."[82] "The words 'I love you' appear more than any other phrase in this book," she admits: "I love you for all of your splashing." "I will love you more whether time moves forward or backwards." "There are at least three ways to love you"; "like this love ever flowing"; "I will love you still"; "I love you."[83] In most of the cases this chapter has addressed, "I love you" is the last thing expressed in language before crossing the threshold into something else entirely. If *saying* "I love you" marks the threshold of language and perhaps even of a certain kind of communicability, then why do we say it? We say "I love you" as a farewell to our human side and as an opening to a kind of communication we cannot properly name.

To be somewhat schematic about it, this chapter has explored three kinds of love humans have for animals. First, there is the erotic—a vexing topic, and one that I have trouble fully pursuing. Does my hesitance imply an anthropocentric moralism, which Singer would undoubtedly accuse me of having? Perhaps. (Online research on the topic takes one to some truly troubling places.) But I have nonetheless devoted substantial space to the topic. My tentative conclusion is that, based on everything we know about ourselves as humans, and especially our capacity for destruction and violence, sexual relations with animals will almost always be non-consensual and problematic. An author like Grebowicz may wish to interpret the relationship between Margaret Howe and Peter as a queer kinship (she goes so far as to read it in relation to Foucault's "Friendship as a Way of Life"!), but such interpretations are premature, failing as they do to consider the terrible violence inflicted upon animals, especially animals held in captivity, like Peter.[84] (I return to this issue in the epilogue.)

A second kind of love is the one described by Haraway and others. This is love in the quotidian, love as a verb, love as entanglement. Being together requires patience and, as Chiang emphasizes, effort. Effort might sound like a simple thing, too simple to use in a book such as this. But, if we are honest with ourselves, everyone knows that sustained effort is one of the most challenging things in the world.

Third, and finally, is the type of love crystalized through *expressions* of love by humans about and to animals: "I miss Tigger a lot. I really loved him"; "I love this bitch beyond words"; "I love you." This kind of love is what Roland Barthes would call a lover's discourse. A lover's discourse is not a discourse about love, it does not describe its conditions and contours. Rather, a lover's discourse is one that simulates love itself. Barthes writes,

> *I-love-you* is without nuance. It suppresses explanations, adjustments, degrees, scruples. In a way—exorbitant paradox of language—to say *I-love-you* is to proceed as if there were no theater of speech, and this word is always true (has no other referent than its utterance: it is a performative).
>
> *I-love-you* has no "elsewhere"—it is the word of the (maternal, amorous) dyad; in it, no distance, no distortion will split the sign; it is the metaphor of nothing *else*.[85]

Is it not inappropriate to apply Barthes' ideas—ideas generated from close readings of Nietzsche, Lacan, Hölderlin, and Proust—is it not inappropriate and *in poor taste* to apply these ideas to how ordinary people speak to their pets? "I love you," I used to say to my cat, Tigger, when he was alive. Was this *I-love-you* Barthes' *I-love-you*, without nuance, without elsewhere, a metaphor of nothing else? Perhaps there is something risky about mixing these discourses, about thinking that Barthes' *I-love-you* is my *I-love-you*. But all the cases I've written about in this chapter are precisely about a certain disrespect for borders, a desire to trespass limits.[86] And because we share the earth with animals, it's not surprising that borders become fuzzy at times.

Love may seem like a weird excess of interspecies relations, something strange that *happens to happen* when a parrot researcher spends too many years with her subject, or when a person feeds her dog every day over an extended period. But if love is a supplement, it is, paradoxically, an essential supplement; for love—in all its ambiguity, all its complexity, all its generosity and violence—is a powerful force in our lives with animals. Barthes again:

> *I-love-you* is . . . neither quite what is uttered . . . nor quite the uttering itself. We might call it a *proffering*, which has no scientific place: *I-love-you* belongs neither in the realm of linguistics nor in that of semiology. Its occasion (the point of departure for speaking it) would be, rather, Music. In the manner of what happens in singing, in the proffering of *I-love-you*, desire is neither repressed (as in what is uttered) nor recognized (where we did not expect it: as in the uttering itself) but simply: released, as an orgasm.[87]

Love, desire, sex, music: a constellation that haunts our relationships with animals. Try as we might to contain it, communication is never just communication. We attempt to communicate with animals (to get animals to do what we want, to teach them how to speak like we do, to let them know how we are feeling, and so on), but we also fear, love, and yearn for them.

If indeed our relationships with animals often end up being more about love (or something like it) than communication, are we doomed to a kind of communicative solipsism of the species? I've already suggested that we are not, and for two reasons. The first reason is that humans can communicate with animals on registers other than the symbolic. The second reason is that love and intimacy are part of our non-symbolic relations with our animal companions. In other words, rather than seeing love as an inconvenient by-product that gets in the way of smooth semantic exchange, we should view it as a supplement that invites association. To fully understand non-symbolic relations between human and non-human animals, we need to embrace rather than exclude aspects like love and intimacy.

In his history of communication from Plato's *Phaedrus* to twentieth-century mass media, John Durham Peters argues that there can be no communication without love. He writes, "I take eros to be the attractions and repulsions between bodies, sexual attraction being only one critical part. I follow Plato in understanding eros as the force fields between bodies that do not touch; sex has to do with bodies that touch in specific ways. The intellectual history of 'communication' is a record of the erotic complications of modern life."[88] Communication as erotic complication: this definition applies as well to Plato as it does do Pepperberg. And yet, as I've emphasized throughout this book, communication is eros with semiosis, attraction and repulsion along with interpretation. Both matter.

There is, moreover, another option for those wishing to communicate with other beings: why limit yourself to the oceans or the trees when there is a whole universe out there? Isn't it possible that the entire universe is teaming with creatures as loquacious, if not necessarily as loving, as us? Questions such as these animate Chapter 3.

II

Alien

Prefatory Remark to the Second Part

A is for *animal,* but A is also for *alien.*

The first part of this book is concerned with animals here on Earth. In the second part we turn our gaze upward—or, better stated, we cock our ears toward the heavens. In the next two chapters, aliens rather than animals are the subject matter. But—someone like John Lilly may query—What is an animal if not an alien intelligence? And, conversely, as a number of writers since the eighteenth-century have asked, What is an alien if not an animal in outer space?

The European Enlightenment bequeathed to us a sharp or even irreconcilable separation between humans and nature. This inheritance essentially means that while humans share material attributes with the non-human world, humans alone experience interiority and volition.[1] But that particular legacy of European modernity has been undermined at every turn. In this book, I have focused on one particular undermining impulse: to trespass the border between the human and non-human, to enter into communion with a being who is like us but is not us.

Within a conceptual paradigm that struggles to properly situate the human, Enlightenment thought has long reckoned with a host of liminal figures that Michael Schetsche and his collaborators call "maximal strangers."[2] Aristotle's question regarding the difference between human and animal was asked again in a new way at the birth of modernity. But the (non-human) animal makes up only one axis of a bifurcated cartography; a second axis concerns the more mysterious and hypothetical figure of the extraterrestrial being living somewhere other than on planet Earth. Although it is hardly a unanimous position, since the eighteenth century a surprisingly large portion of writers have affirmed the existence of extraterrestrial life.[3]

Enlightenment conceptualizations of human and non-human categories were increasingly materialized in the second half of the twentieth century. Following the Second World War, scientists embarked upon the deep exploration of oceans and cosmic space—two realms previously inaccessible to the senses and to empirical investigation. As quintessential frontiers of the postwar period, the "outer" space of the cosmos and the "inner" space of oceans were conceptualized as structurally isomorphic twins, subject to the same method of scrutiny.

The scientific exploration of these frontiers was haunted from the outset by the specter of as yet unknown, alien, and intelligent forms of life. These intelligent life-forms—shadowy as they were—moved from a topic of speculative thought to the subject of empirical investigation. During the Cold War, nations on both sides of the Iron Curtain funded research into these newly opened realms. But those research initiatives probably did more to bolster fantasy than to tame or attenuate it. Scientific discourse became irretrievably entangled with science fiction and even, at times, with conspiracy theory. Authors like Carl Sagan and artifacts like Voyager's Golden Record blur the boundary between scientific rigor and fantasy bordering on delirium. Such fantastic investments touched the oceans no less than the heavens.

A host of new maximal strangers came into view—or, perhaps, into earshot. Armed with new knowledge and new technological capacities, scientists posed a series of novel questions:

Might it be possible not only to examine, but to enter into communion with, creatures very unlike ourselves?

How might we reach these maximal strangers?

What sounds do they produce?

Sound and listening are crucial elements in this terrain. Pythagoras listened for the music of the spheres, and if today we dismiss his position as pre-Copernican superstition, we have nonetheless not stopped listening. Breakthrough Listen—"the largest ever scientific research program aimed at finding evidence of civilizations beyond Earth"—was launched in our own century, in 2015. According to the website, the instruments that Breakthrough Listen uses "are sensitive enough to hear a common aircraft radar transmitting to us from any of the 1,000 nearest stars."[4] Our ears, it seems, have become much sharper since the time of the ancients, and more high-tech.

Chapters 3 and 4 follow materially from the discussions of animals in the previous two chapters. The turn from animals to aliens is no mere whimsy on my part. Rather, as I show in what follows, animals and aliens pose similar problems for thought.

✷ 3 ✷

The Incomparable

Sometime around 1966, John Lilly wrote a letter to the astronomer Frank Drake with some pointed criticisms about a recent paper by the latter.[1] Today, Drake is best known for the "Drake equation," an attempt to estimate the number of intelligent extraterrestrial civilizations in the Milky Way. Although the equation was highly conjectural even by Drake's own admission, it did suggest that, at least on the face of it, there should be *many* intelligent civilizations in the galaxy and not just one—our own. In Green Bank, West Virginia, in 1960, Drake launched Project Ozma, which sought to detect radio transmissions from two stars approximately the same age as the sun and lying roughly sixty-six trillion miles from Earth. "Detection," here, meant scanning for regular or repeated signals, which, hypothetically at least, should stand out from the background of otherwise random cosmic noise. Drake's assumption was that any such signal should be attributed to the intentional act of an intelligent civilization.

Scribbling in the margins of Drake's latest attempt at formulating our chances for extraterrestrial communion, Lilly wrote as follows:

> Dear Frank,
>
> Fine presentation—but there's a "missing link"—(the one I keep raising): conceive of any other intelligent communication form (ICF) (alive or not): conceive of an infinity of metatheoretical views of "reality external" and, simultaneously, "reality internal" including the "thinking software itself": can we, with our particular stage of development even conceive of proper means of communication with any of such different "computers"? ICF (≥ our own) ∴ may not "signals" already exist from a very large number of grouped old alien IFCs which we label "cosmic noise"?

Lilly goes on to propose that, rather than searching for what to us seems like a patterned signal that would suggest the presence of an intelligent communication form (ICF), we should instead assume the existence of ICFs and then try to sort through the seeming randomness to find deliberately created complex patterns. The problem, Lilly suggests, is not that we are looking in the wrong place or that we are limited by the current state of our technology, but rather that we are "pattern recognition and metatheory limited."[2]

Lilly's hypothesis should seem familiar, as it is exactly the same one he applied to dolphins. In both cases—dolphins and aliens—Lilly proposed beginning with the assumption that they are trying to communication with us, and then doing everything possible to find out what they are saying. The alien/dolphin analogy is no accident; rather, it was fundamental to thinking about interspecies communication in Lilly's time. In the 1960s and '70s, Cold War science created a structural isomorphism between outer space and the "inner space" of oceans.[3] Outer space and inner space were conceptualized as two new frontiers of knowledge, as two realms that had previously been inaccessible to empirical inquiry, but that had finally become open to scrutiny with the development of new technologies.

All of Lilly's posturing belied his anxious feeling that we may, in fact, be radically alone. The problem that Lilly faced, and indeed the problem that we face today, is explaining the singular place of the human in the cosmos. This is an old problem, even an ancient one, as this chapter will show. But we seem not to have made any real progress on the question. On the contrary, with every scientific revolution, the problem of humanity's singularity appears more intractable, more perplexing. Since Copernicus, we know that we do not reside at the center of the universe. And since Darwin, we know that we are biologically evolved creatures like other life-forms. And yet, human exceptionality is a core assumption of modernity—even if that assumption has recently been called into question in certain quarters.

Considering the vast depths of the cosmos, Blaise Pascal (1632–1662) bemoaned "the eternal silence of these infinite spaces."[4] In the twentieth century, writers like Loren Eisely continued to ponder humanity's tremendous loneliness in the universe. This is how Eisley describes our predicament:

> There is nothing more alone in the universe than man. He is alone because he has the intellectual capacity to know that he is separated by a vast gulf of social memory and experiment from the lives of his animal associates. He has entered into the strange world of history, of social and intellectual change, while his brothers of the field and forest remain subject

> to the invisible laws of biological evolution. . . . When we were children we wanted to talk to animals and struggled to understand why this was impossible. Slowly we gave up the attempt as we grew into the solitary world of human adulthood; the rabbit was left on the lawn, the dog was relegated to his kennel.[5]

This chapter deals primarily with the kind of thinking born from assuming human exceptionalism—that is, from relegating the dog to the kennel. To be sure, this sense of cosmic loneliness implies a particular positioning of humanity. While our place in the cosmos may, in a sense, have always been registered as a kind of dilemma, the sense of crushing loneliness and isolation—the existential dread of being mere specks—is not a sentiment shared by all people everywhere. The animal lovers discussed in the previous chapter would balk at Eisley's characterization of adulthood as the time of life when dogs are banished to the yard. Even so, those individuals embracing a splendid universe of shared animal habitation (Donna Haraway, for example), would be the first to acknowledge their human difference from animal kin. For someone like Haraway, this difference leads not to loneliness but to joy, responsibility, challenges, and love. But it is a difference, nonetheless—one you can mitigate or downplay but never overcome.

In the three centuries since Pascal, many have proclaimed that making contact with an intelligent alien species would be the most significant event in human history. But with every failed attempt we are thrown back on ourselves. How to understand our potentially singular position in the universe? What are the real possibilities for verifying it one way or another? If, as many in the know suspect, we may soon confirm the existence of life (if not "intelligent life") on other planets, what do we do in the meantime?[6] What would such a confirmation mean for us as a species? And what does the absence of confirmation mean right now?

Although John Lilly claimed that his investigations into dolphin intelligence were a dress rehearsal for future extraterrestrial contact, the opposite was truer: he frequently turned to extraterrestrial speculation when frustrated by his inability to break through the human/dolphin communication barrier. In other words, although he described dolphins as easily accessible conversation partners, those creatures eluded him for his entire life, generating all kinds of fantasy and whimsy, from bizarre communication gadgets and experiments with hallucinogenic drugs, to forays into music theory (and metaphysics) and speculations about alien intelligence. Enrico Fermi wondered why, if there are millions of exoplanets in the universe, we have not heard from aliens—this is the famous "Fermi paradox," which I return to later in the chapter. For his part, Lilly wanted to know why the various large-brained mammals of the world have not tried to engage each other in conversation. Lilly wasn't alone asking this question: in the 1960s and '70s, scientists around the world oscillated between animal and alien communication partners, turning to animals when questions about alien intelligence reached an impasse, and vice versa.

To some degree, the frustrations, failures, and occasional glimmers of success in human/animal communication generated questions about unknown and potentially unknowable intelligent beings elsewhere in the universe. The alien, in this sense, holds a similar position in the conceptual cartography of interspecies communication as music or love, because it enters the picture when attempts at terrestrial cross-species communion are exhausted. In many instances, these figures of communicative excess function together—for example, when whales are associated simultaneously with music, generosity, and cosmic tranquility.

As we will see in the pages that follow, listening often becomes a key technique and metaphor for those interested in alien encounters. Pascal, after all, spoke of the universe's eternal *silence*—a silence, presumably, that can only be broken by chatter. Listening, silence, noise, sound, speech, music, and other sonic figures saturate discourse on alien contact. We will have occasion in this chapter to examine many of those figures.

This chapter tackles existential questions, but these questions are also, as we have noticed, ones that touch on the most personal emotions of all—melancholy, love, loneliness, and desire. In other words, although dealing ostensibly with the unknown, the riddle of extraterrestrial life is registered to us here on Earth as a defining feature of who and what we are. The unknowable returns to us as direct emotion, a paradoxical situation giving rise to acute feelings of yearning and the sublime (to add to the list of melancholy, love, loneliness, and desire). Commenting on the Voyager's Golden Record (a phonograph record containing music, sound, speech, and images that was sent into space), Carl Sagan opined, "It is, as much as the sounds of any baleen whale, a love song cast upon the vastness of the deep."[7] The Golden Record contained a recording of whale song, but it, too, according to Sagan, was a song—a love song, no less. But it is a love song with an unknown addressee, a love song in search of someone or something to love.

A Telescopic View of the Extraterrestrial Life Debate

Those new to the subject will be surprised by how often people have affirmed the existence of extraterrestrial life. To the Epicureans of antiquity, for example, the existence of alien life was a given. As Epicurus himself wrote, "There are infinite worlds both like and unlike this world of ours," and, later in the same text, "We must believe that in all worlds there are living creatures and plants and other things we see in this world."[8] The assumption of multiple inhabited worlds came to be known as *pluralism*, which became a key term in many of the subsequent debates.

Until the early modern period, pluralism remained something of an outlier theory, meeting resistance from Platonism, Aristotelianism, and church theology. (The history of the extraterrestrial life debate has been documented in stunning detail by Michael J. Crowe, Steven J. Dick, and Karl S. Guthke, among others.)[9] It was only in the seventeenth century that pluralism gained widespread traction. While the first half of that century was a particularly tumultuous period in the extraterrestrial life debate,[10] the ideas of Descartes and his contemporaries laid the groundwork for what became the most popular pluralist work in history: Bernard le Bovier de Fontenelle's *Conversations on the Plurality of Worlds*.[11] Although it drew on cutting-edge scientific knowledge, it is not a work of science. It is, rather, a popular book written in the form of a dialogue between an elderly philosopher and a marquise; as such, it does not require Fontenelle himself to explicitly declare pluralist views. Nonetheless, by most accounts, Fontenelle's *Conversations* ushered in the heyday of pluralism, and by the beginning of the eighteenth century, many European scientists and writers openly declared their belief in living beings populating other worlds.

More than those before him, Fontenelle's thinking was shaped by the particular form of anthropological thinking born from European expansion. When, for example, the characters in *Conversations* describe inhabitants of Venus as "little dark, sun-burnt people, scorched with the Sun; full of wit and animation, always in love, always making verses, listening to music, having galas, dances, and tournaments," they repeat horrible colonial clichés.[12] Fontenelle directly imports racist caricatures about Africans to describe beings on Venus.

The comparison between extraterrestrials and colonized peoples is fundamental to *Conversations*. In this book's introduction, I quoted this passage from *Conversations*: "Oh no, I don't believe there are men in the moon. We see how much all nature has changed even when we have traveled from here to China; different faces; different figures; different manners; and almost a different sort of understandings: from here to the moon the alteration must be considerably greater. When adventurers explore unknown countries, the inhabitants they find are scarcely human; they are animals in the shape of men, even in that respect sometimes imperfect; but almost devoid of human reason."[13] If this horrifically (and characteristically) racist passage gives the impression that Fontenelle believed in a kind of "spectrum" of reason, a gradation between (White European) men and savage Others, this could not be further from the truth. Throughout the book, Fontenelle insists on a categorical distinction between men (or people, *hommes*) and mere inhabitants (*habitants*). In an attempt to clear up any confusion, he writes in the preface, "The posterity of Adam cannot have colonized the moon; therefore the inhabitants of that planet are not descendants of our first parents; it would be a difficult point in theology to account for the existence of men who had any other ancestor."[14]

By contrast, Leibniz speculated that if we met a lunarian, we would likely grant him "the title man [*d'homme*]." Even so, Leibniz continued, because the lunarian could not have descended from Adam, his baptism would be a highly controversial affair.[15] Other writers wondered who, in the absence of Christ, would redeem the inhabitants of distant worlds from their sins? Extraterrestrial baptism was a recurring theme in the medieval and early modern periods, leading one historian to refer to the problem of the "planet-hopping Jesus."[16] As was the case with Fontenelle, who transferred colonial tropes about tropical Africans to the inhabitants of Venus, the theological debate about whether aliens have souls was firmly rooted in ongoing conversations about the potential for European Christians to save the souls of New World natives.[17] In the colonial context, as well, controversy raged over the question of whether all people, including "Indians," "derived from one single Adamic race."[18]

While most authors speculated that extraterrestrials are probably spiritually degenerate beings with humanlike bodies, in the early modern period, the centrality of the soul to the extraterrestrial life debate began to recede. The debate was increasingly defined by two somewhat opposing hypotheses. First, by the turn of the eighteenth century, most European writers accepted the probability of life, even intelligent or rational life, on other planets. A second central feature of the debate during this period was that we, as humans, can have no positive knowledge about those life forms. (The ideas of some eighteenth-century writers resemble Epicureanism on this point.) On this score, it is instructive to compare the views of two key Enlightenment figures: John Locke (1632–1704) and Immanuel Kant (1724–1804). Both Locke and Kant believed in the probability of extraterrestrial intelligent life, although Kant grew cautious in his later writings. But Locke and Kant drew different conclusions from our inability to attain positive knowledge of those beings.

As an empiricist, Locke held that ideas are the results of the senses, which also means, he thought, that our ideas are limited by the capacity for sensation. Locke could only muse about the ideas that "creatures in other parts of the universe may have, by the assistance of senses and faculties more, or perfecter than we have, or different from ours."[19] We have no way of knowing, he wrote, the "several sorts of vegetables, animals, and intellectual corporeal beings, infinitely different from those of our little spot of earth," that likely populate the universe. Such ideas are "not for us to determine." What is important, rather, is admitting that they must and do indeed exist. To say that they do not, Locke says, is like a blind person saying that colors do not exist, or like a mole denying the ideas an eagle derives from its quicksightedness. Only an arrogant and unchristian person would dare to imagine that he might access the ideas of a creature with a fundamentally different body and sensorium; and it would require an equal measure of arrogance and lack of Christian faith to deny the truth of an extraterrestrial's sensory reality, however different it may be.[20]

At least in his early work, Kant assumed the existence of extraterrestrial intelligence, and he returned to speculations on the topic at critical moments throughout his life.[21] For example, when attempting to define the character of humanity in *Anthropology from a Pragmatic Point of View*, he wrote,

> The highest species concept may be that of a *terrestrial* rational being; however, we shall not be able to name its character because we have no knowledge of the *non-terrestrial* beings that would enable us to indicate their characteristic property and so to characterize this terrestrial being among rational beings in general. It seems, therefore, that the problem of indicating the character of the human species is absolutely insoluble [*schlechterdings unauflöslich*], because the solution would have to be made through experience by means of the comparison of two *species* of rational being, but experience does not offer us this possibility.[22]

Kant points to what he sees as the inherent difficulty of defining the human. Because we have nothing to compare ourselves to, he says, it is not possible to determine what defines us. The philosopher Peter Szendy comments that according to the Kantian dilemma, humanity can be defined "only by undefining itself in a movement of *comparison without a comparative term*."[23] Trying to define ourselves through comparison without anything to compare ourselves to can only result in undefining, in insolubility. The alien is, for Kant, the placeholder for a comparison that could exist but remains forever out of reach. The insolubility that Kant identifies here will continue to haunt the extraterrestrial life debate. Indeed, in our time a key question remains: How do we search for life on other planets when we have only life on Earth as a basis from which to begin?[24]

In broad strokes, writers in the seventeenth and eighteenth centuries affirmed the existence of diverse life forms on other planets but also remarked on our inability to ascertain positive knowledge of those life forms. Over the course of the next century, new factors were entered into the equation, some of which purported to offer the possibility of positive knowledge. One aspect anticipated the work of Frank Drake and others a century later: in the mid-nineteenth century, a small group of individuals created proposals for direct communication with beings on other planets. Charles Cros (1842–1888) was perhaps the first to put forward concrete plans for communication with extraterrestrials.[25] Cros devised a method of focusing light through parabolic mirrors as a way to make our presence known, and even to send messages through regular flashes of light. The success of the project will "be a moment of joy and pride," he wrote. "The eternal isolation of the spheres is vanquished."[26]

In order to better understand this episode, it may help to remember that this was the same Charles Cros who had proposed an early mechanism for sound reproduction. In April 1877, he sent a sealed letter to the French Academy of Sciences with his idea, which essentially consisted of reversing the process of imprinting sound on a surface, which was by then well established. In January of the next year, Thomas Edison patented the phonograph, which worked by a process similar to the one described by Cros. Because Cros did not make a working model of the phonograph, the argument that he had invented a sound-recording device several months before Edison held no water in any legal system of the time. Even so, according to one recent commentator, "a bitter debate erupted in the subsequent months about who deserved recognition for its invention."[27]

In a more sophisticated historical account than one that merely trades in firsts, Jonathan Sterne has shown how important transduction was in the mid-nineteenth century—transduction, here, refers to the transformation of energy from one form or medium to another. That sound could be transformed between media was already well known by Cros' time: Chladni showed nearly a century earlier that sound vibrations could form patterns in grains of sands on a flat surface; Koenig showed that fluctuations in sound could affect flames; and—perhaps most importantly for the current discussion—Bell and Blake showed that, when passing through the eardrum of an eviscerated human cadaver, sound could stimulate a stylus to trace patterns on a piece of paper or glass. Sound transduction compelled a reevaluation of hearing. This reevaluation was the cause, rather than the result, of an invention like the phonograph, and so, as Sterne says, although Cros did not invent the phonograph, he, like many others at the time, "grasped the principle behind the phonograph before it was actually invented."[28]

Cros and his circle of friends in Paris probably experienced equal parts frustration and vindication when the phonograph proved to be an astonishing success. In Cros' own mind, the episode must have seemed uncannily similar to his proposals for interplanetary communication, which, again, he could not implement, but in which case, again, he had understood an underlying principle. If the ability to record and play back a human voice was greeted with amazement by the public (even, that is, when the underlying principle had long been clear) then surely the public would feel the same, at some later date, about our communion with extraterrestrials.

Cros was among the first to advance a system for alien contact, but he was by no means the last. A proposal for interplanetary communication via electrical signal by Nikolas Tesla followed, and in 1921 Guglielmo Marconi announced that he had intercepted wireless messages from Mars.[29] These views were by no means accepted by the scientific community or by the public, and, in the most general terms, by the end of the nineteenth century "pluralism had taken on an attenuated form, at least if compared with the pluralism of midcentury."[30] But even the remote possibility of making contact with extraterrestrials challenged Kant's notion regarding the insoluble problem of defining the character of humanity. What if that character is soluble after all? The problem was (and is) that any attempt to communicate with or receive signals from an extraterrestrial being must begin with a set of assumptions about what that kind of being is, and those assumptions can only ever be based on what we know of the one species capable of sending and receiving signals: us. Although what later came to be known as SETI aims to overcome humanity's "insoluble" character, it arguably begins with and is mired in that insolubility.

Astronomers had for centuries promulgated the existence of human-like or otherwise "intelligent" life on Mars, Venus, the Moon, and even the Sun. By the second decade of the twentieth century, though, scientists tended to emphasize that if life does exist elsewhere, it would probably be rare and very far away. As the British astronomer James Jeans wrote in 1929, "Life must be limited to a tiny fraction of the universe." For Jeans, "a tiny fraction" obviously meant something very small, and of course it is (or at least *was* in comparison with many people writing just a decade earlier, for whom contact with Martians was imminent, if it had not happened already). In the following decades, however, a "tiny fraction of the universe" was recognized as a very large amount, at least in absolute terms. For example, and according to the most cutting-edge thinking in the twenty-first century, to say that the chance of an intelligent civilization evolving on another planet is one in a trillion means that, over the course of the history of the universe, ten billion such civilizations have existed.[31]

Something like this position had already been reached by the second half of the twentieth century, at the beginning of the Cold War Space Race, as the first rockets were launched on both sides of the Iron Curtain. Having jettisoned fanciful notions of Martian invaders, most sober-minded scientists hypothesized that many civilizations would arise in a universe of such vast proportions, even if life were scarce.[32] If this is so, the physicist Enrico Fermi is reputed to have asked, then "Where is everybody?" Fermi allegedly made this remark casually during a lunchtime conversation with some colleagues around 1950, and while Fermi's exact words are perhaps lost forever, the gist of his question has since gained something of a mythical status in the extraterrestrial life debate. Today, that question is typically formulated as a paradox (sometimes referred to as the "Fermi paradox"): If the universe is likely populated with many civilizations, then it stands to reason that we should by now have had some experience with at least one of them. But we have not. Why?

The underlying assumption of the Fermi paradox—namely, that we *should* have heard from another civilization, or perhaps, that we should be hearing from many all the time—sounds a strident rejoinder to Pascal's "the eternal silence of these infinite spaces frightens me." If Pascal is usually understood as an anti-pluralist whose terror stems from a crushing loneliness, then Fermi's question raises the uncanny specter that we are not alone even though we appear to be. Attempts to resolve the Fermi paradox are legion, and come from many quarters: reputable academics, amateur enthusiast bloggers, tech entrepreneurs, science fiction authors, and conspiracy theorists. The line between these categories is often blurry.

One approach to resolving Fermi's paradox is known as the Great Filter.[33] This theory begins with the assumption that intelligent life evolves in stages: from dead matter, to living organisms, to intelligent living organisms, to technologically equipped beings that "explode" out into the universe.[34] But contrary to those who crunch the numbers and conclude that if there are *enough* Earthlike planets, then at least *some* will develop humanlike beings, the Great Filter theory proposes that at some specific point along the chain there must be a hurdle that filters out almost all development. In other words, if for Drake there is just one long and smooth funnel, Great Filter proponents imagine a funnel that is clogged up at one point.

But at what point, exactly? If the Great Filter is located between inorganic and organic life, for example, then perhaps the Earth is exceptional, and the emergence here of life was radically improbable. Or perhaps the Great Filter happens later, almost always preventing single-cell life from developing into multicellular life. In both cases—inorganic to organic, and single- to multicellular—the Great Filter is behind us, which means that complex animals (including humans) must have somehow, almost miraculously passed beyond it. But what if the Great Filter is not behind us, but ahead of us, in the future? In this version of the theory, intelligent beings with technological capacity may be a dime a dozen in the universe, and the fact that they have not found a way to communicate with one another might mean that *that* stage is precisely where the Great Filter is located. In other words, we are possibly at the precipice of our own disastrous demise.

One astronomer pithily summarized this version of the Great Filter theory as follows: "Throughout the universe a civilization's own technological advancement overwhelmingly predicts its destruction."[35] Few alive today would be surprised to learn that humans are at the threshold of a Great Filter, and that before anything so grand as "making contact" with extraterrestrials happens, humanity will kill itself, will go extinct. Millions, maybe billions of societies throughout the universe may have reached the point where we are now, a point where dreams of interplanetary communication are possible, where terraforming other planets seems viable. But perhaps this is as far as it goes before, for some reason, all such civilizations are filtered out of existence.

Searching for Extraterrestrials

The historical backdrop provided in the previous section undermines the common conception that the search for extraterrestrial intelligence (SETI) began with a theoretical statement by Cornell astronomers Giuseppe Cocconi and Philip Morrison in 1959 and was first implemented two years later by Frank Drake in Green Bank, West Virginia. Nonetheless, Drake's efforts crystallized a lot of earlier thought on the topic, which, as we have seen, goes back several centuries—at least to the time of Charles Cros, if not earlier.

In 1961, ten people (all men, and all White except for Chinese-American astronomer Su-Shu Huang) gathered in Green Bank for a symposium on SETI. Drake, Philip Morrison, Carl Sagan, and John Lilly were all in attendance. They called themselves the "The Order of the Dolphin" in honor of Lilly's work.[36] Of course, Drake's Project Ozma never detected an alien signal. SETI was essentially shelved for the rest of the 1960s, since, in light of Cold War priorities, the expense simply could not be justified. SETI researchers have continually struggled to acquire NASA funding ever since—between 1970 and 1993, the agency has never devoted more than 0.1% of its annual budget to SETI projects.[37] In most cases, NASA support was more symbolic than anything else. For example, when NASA introduced one of its admittedly meager SETI attempts in 1992, it chose Columbus Day for the launch, and the fact that the launch marked the anniversary of Columbus's arrival in America in 1492 was not lost on those involved.[38]

Other SETI endeavors in the second half of the twentieth century were supported by an improvised combination of university funding, private funding, and volunteerism. A good example is the long-running project Big Ear, which made use of a radio telescope at The Ohio State University that had been decommissioned and practically abandoned. Big Ear searched continuously from 1973 to 1995 with volunteer assistance from students and professors at the university. An example of privately funded SETI research is the SETI Institute in Mountain View, California, which was founded in 1984.[39]

In the absence of another intelligent species with which to compare humans, SETI researchers have been compelled to work at a general register, asking, What is the most efficacious strategy for one civilization attempting to communicate with another? Tesla had already proposed in 1901 that interplanetary communication would undoubtedly take place through numbers: "Certitude as to the receipt and interchange of messages would be reached as soon as we could respond with the number 'four,' say, in reply to the signal 'one, two, three.'"[40] But how would these numbers become perceivable by a distant civilization? As Daniel Chua and Alexander Rehding write, "We may share a 'language' of numbers with the alien, but if we do not interface via a physical medium to transduce the data, there is no contact; the numbers will remain abstract and immobile. . . . For transmission to occur, numbers need to materialize and move in time."[41]

Recall project "Big Ear": one is struck, when surveying the history of twentieth-century SETI, by the ubiquity of aural metaphors. The technological basis of SETI is radio astronomy, which is concerned with electromagnetic waves that obviously cannot be heard, and that must be transformed into sound waves to become audible to the human listener. Radio waves are instances of electromagnetic radiation at frequencies lower than microwaves. Radio waves travel at the speed of light in a vacuum and only slightly slower in planetary atmospheres. They are emitted by many kinds of astronomical objects, a fact known since the 1930s. Why, then, the common recourse to listening? To some extent, the choice of listening and the ear is attributable to the fact that scientists often modulate electromagnetic waves into soundwaves to be heard. Some consider the ear the preferred sense organ for the detection and monitoring of patterns, for example with sonar, heart monitors, and Geiger counters. The idea here is that it is easier to listen for patterns or irregularities without fatigue than it would be for the eye and visual representation. *Sonifying* electromagnetic waves—that is, transforming them into sound, takes advantage of the ear's "ability to detect anomalies that might pop out of a continuous stream of information."[42]

But we should not be so quick to conclude that the ear has an essential advantage over the eye in such matters. After all, when, in the early 1930s, a young engineer at Bell Laboratories named Karl Jansky accidentally discovered radio waves emanating from beyond Earth for the first time in history, his data was collected visually through an analog paper and pen system. Jansky, of course, did not capture signals from an alien civilization; rather, he detected radio waves from the densest region of the Milky Way, which scientists now believe to be a black hole. If Jansky and other, earlier radio astronomers used visual rather than audio representation, then recourse to aural metaphors in subsequent decades must have some other foundation than the efficiency of the ear. Indeed, frequent invocations of listening to space seem to be based rather on ideologies of aurality that freight listening and the ear with metaphysical significance. Science fiction often capitalizes on the powerful image of a solitary scientist listening to the skies. In many such representations, the scientist is cocooned by headphones that are connected to a giant radio telescope pointed toward the heavens. What is this radio telescope other than a Big Ear, a massive aural prosthesis channeling alien signals into the scientist's ears below? Consider for example, the following description by the protagonist, Ye, from Liu Cixin's already classic *The Three-Body Problem*: "The search for extraterrestrial intelligence is a unique discipline. It has a profound influence on the researcher's perspective on life. . . . In the dead of night, I could hear in my headphones the lifeless noise of the universe. The noise was faint but constant, more eternal than the stars. Sometimes I thought it sounded like the endless winter winds of the Greater Khingan Mountains. I felt so cold then, and the loneliness was indescribable."[43]

When Ye does, for the first time in human history, detect an alien signal, she does so not through the ear but with the eye. Even so, Liu deploys a set of mixed metaphors typical in SETI descriptions:

> One night, Ye was working the night shift. This was the loneliest time. In the deep silence of midnight, the universe revealed itself to its listeners as a vast desolation. What Ye disliked most was seeing the waves that slowly crawled across the display, a visual a record of the meaningless noise Read Coast picked up from space. . . . On this day however, Ye saw something odd when she glanced at the waveform display. Even experts had a hard time telling with the naked eye whether a waveform carried information. But Ye was so familiar with the noise of the universe that she could tell that the wave that now moved in front of her eyes had something extra. The thin curve, rising and falling, seemed to possess a soul.[44]

Here, Ye uses her eyes to detect regularities in the waveform display, but the situation is still described in terms of silence and noise. A similar mixed metaphor can be found in an ambitious SETI report from 1971. Somewhat ironically titled *Project Cyclops,* a reference to the one-eyed giants of mythology described by Hesiod and Homer, the report nonetheless turns to aurality at critical moments. For example, when outlining strategies with the greatest potential for detecting extraterrestrial life, the report states, "Nature has provided us with a rather narrow quiet band in this best part of the spectrum that seems especially marked for interstellar contact. It lies between the spectral lines of hydrogen (1420 MHz) and the hydroxyl radical (1662 MHz). Standing like the Om and Um on either side of a gate, these two emissions of the disassociation products of water beckon all water-based life to search for its kind at the age-old meeting place of all species: the water hole."[45]

Invoking the sacred Hindu Om again illustrates the way that SETI research harnesses sound as metaphysics. The astronomer Paul Horowitz offers a less extravagant, if nonetheless rich, sonic metaphor. Commenting on the fact that communication via radio waves is efficient, he notes that "radio noise backgrounds are extremely quiet. You can hear a pin drop in the radio bands, and in the universe."[46] His description is redolent of the one found in James Gunn's 1972 science fiction novel *The Listeners,* when the protagonist hears a message from an alien civilization as short bursts of sound piercing through a universal background of static noise. In many ways, SETI researchers and science-fiction authors have mobilized metaphysical notions of listening in the service of extraterrestrial communication. At least in European thought (although the same could possibly be said of many other cultures as well), sound and music have historically been understood as ways of reaching the beyond: beyond life, beyond the present, beyond the human, and beyond Earth.[47]

When reading about astronomy, I am often reminded of the advent of contemplative listening in nineteenth-century Europe: for the Romantics, the paragon of a music listener is someone who listens with eyes closed to great symphonic works that transcend time and place. The image of the nineteenth-century music listener—eyes lowered, listening in concentrated stillness—is transformed into the scientist sitting in a dark room "listening" to the universe beyond. This experience, such as it is, says more about the way that European ideology has shaped listening as a practice of focused and close attention than it does about astronomy or space.

The scare quotes around the word *listening* in the previous paragraph are important, since, as I have emphasized, one cannot hear radio waves. Having said that, it is possible to understand the practice of listening in less a reductive sense. The philosopher Peter Szendy followers Heidegger's suggestion that hearing and vision do not have their genuine element in sense-perception. He glosses Heidegger as saying that "the partition between the sensible and the nonsensible—between hearing proper and hearing as a purported metaphor for thinking—is really metaphysical."[48] Projects like Big Ear affirm listening in this metaphysical (rather than simply metaphorical) sense. And this explains why the idea of listening for alien signals can be mystifying. Although we know that one cannot hear radio waves, the connotation of massive technological "ears" continues to do substantial ideological work. Should we ever receive such a signal, rest assured that newspapers around the world will proclaim that we have "heard" from aliens from the first time.

Recall Chua's and Rehding's remark that "for transmission to occur, numbers need to materialize and move in time." So there's materialization and movement in time (through sound, for example), but there's also the assumption that before we get to materialization, numbers serve as a useful shared "language" between all the creatures of the universe. This was Tesla's idea, as I've mentioned, but it was also Carl Sagan's, who theorized the conditions for interplanetary communication in a similarly numeric way more than sixty years later. Of particular relevance is his thought experiment published in *Intelligent Life in the Universe.*[49] The thought experiment deliberately leaves out a lot; for example, the fact that, by Sagan's own admission, "a simple query and response to the nearest postulated technical civilization would require periods approaching 1000 years."[50] With all such obstacles acknowledged, Sagan nonetheless speculates on how a message might be sent, received, and understood across the vast expanses of space.

The first issue is one addressed at the beginning of this chapter: that is, how to ascertain that received waves are indeed deliberate messages. As Norbert Wiener put it in the more general terms of information theory, "In the problem of decoding, the most important information which we can possess is the knowledge that the message which we are reading is not gibberish."[51] Sagan, like many proponents of SETI, was not overly concerned with this step, and he assumed that any kind of regularity would be a sign of non-gibberish, of meaning. Nor did Lilly disagree with the step; he simply felt that the kind of "regularity" people were searching for was far too simple. (After all, natural signals detected from celestial bodies may repeat with each rotation, which means that periodicity is no proof of deliberate intentionality. This, I believe, is why Lilly complained that we are "pattern recognition and metatheory limited.")[52]

Sagan asks us to imagine a short message, whose information we record, "perhaps on magnetic tape," slowed down, and then displayed "in some conventional representation as, for example, zeros and ones." A new question arises: not, "Where is everyone?" but, "What are they saying?" Like Lilly, Sagan recognizes the need to cut through many layers. "An advanced technical civilization is trying to communicate with us. . . . They are not likely to speak English or Russian. They have had a different evolutionary history. They are on a planet with perhaps an entirely different physical environment."[53] How, then, might such a civilization proceed? How might a message be sent?

It is an alluring question. Sagan begins reasonably enough by asking how an abstract stream of frequencies might, in the context of two far-flung civilizations, be rendered meaningful. For something to be meaningful, Sagan suggests, it must be *perceptible* (what Chua and Rehding call "materialized"). Sagan therefore crafts his answer to the question of how a message might be sent through a particular interpretation of natural selection. "The number of possible sensory receptors in extraterrestrial organisms is apparently limited," he writes. In environments where water and gas are ubiquitous, sensory organs capable of chemical analysis will be useful, which is to say that organisms evolved in such environments would rely on the senses of taste and smell. The value of a sensory organ capable of perceiving sound (that is, hearing) greatly depends upon a planetary environment, and Sagan doubts the wide distribution of an organ with this capacity. About "pressure receptors" (that is, what we would call touch), he is more optimistic, suggesting that the ability to detect levels of pressure would be widely useful across many different kinds of environments.[54] But it is sight, or to put it in evolutionary terms, the capacity for an organ to register electromagnetic radiation, that Sagan puts his money on. Because most of the star systems that are capable of sustaining life radiate in what we would call the visual spectrum, sensory organs operating in this range will also be probable. "Furthermore," he continues, "the visible part of the spectrum is the wavelength range least likely to be absorbed by atmospheric constituents. Thus, for fundamental physical reasons, the visible wavelength interval is a 'window' region of transparency in all planetary atmospheres."[55]

Based on this argument, Sagan suggests that, faced with the problem of communicating across the cosmos, what one needs is a picture. But how can we send a picture by means of a signal without the necessary information about how to modulate the signal into the picture, and how can we send that necessary information other than through another signal, which the receiving civilization has no way of understanding? Sagan explores one possible answer using a proposal by none other than Frank Drake. Here is the message: see if you can, with everything that has been said, decode it.

11110000101001000011001000000001000001010010000001100101100111100000
01100001101000000000100000100001000010001010100001000000000000000
00000001000100000000000001011000000000000000000000010001110110101101
01000000000000000000000100100001110101010100000000010101010100000
00000111010101011101011000000001000000000000000000010000000000000
01000100111111000001110100000101100000111000000001000000000001000000
00010000000001111100000001011000101110100000000110010111101011111000100
0111110010000000000000111110000000101100011111110000001000000110000000110
000100000110000000011000101001000111100101111

Imagine, for example, that Breakthrough Listen detects this sequence, which takes 20 seconds to complete, and is received every 101 seconds. (In the binary representation above, ones stand for long pulses of an electromagnetic radio wave and zeroes for short pulses.)[56] Assuming that this is a message sent by an intelligent civilization, what does it mean? Following the first SETI symposium in 1961 discussed above, Drake mailed this riddle to the participants. No one was able to solve it. (Don't feel bad if you also were not able to.)

Faced with the above sequence of ones and zeros, a logical place to start might be simply counting how many units are present. The answer is 551.[57] A second question might be how many zeros and how many ones are in the sequence. There are 172 ones and 379 zeros. 379 is a prime number—might that imply that the zeros in the sequence are a possible candidate for coded information? 172, however, has no obvious special qualities, suggesting that the total number of ones and zeros respectively may not hold the key. Does 551 have any particular mathematical properties? Yes, it is a semiprime, meaning that it is divisible only by two prime numbers, 29 and 19. Perhaps we are inching closer.

Recalling that some kind of visual sensory organ is a probable result of natural selection, 551's semiprimeness takes on new meaning. How might the fact that 29 × 19 = 551 (and the fact that 551 is only divisible by 29 and 19) help us to *see* something? From the many times we may have considered renting or buying an apartment, painting a wall, or choosing a pane of glass for a window, one obvious property of two numbers multiplied stands out: ascertaining surface area. This may imply, or at least Drake hoped it would imply, that 29 and 19 are the sizes of a two-dimensional diagram. If we take a "1" (in the actual signal this might be a long pulse) to be a filled-in square and a "0" (in the actual signal this might be a short pulse) to be an empty square, then we can arrange the above sequence of ones and zeros into an image of 29 across and 19 down (or "19 groups of 29 characters," as Drake puts it).[58] The result appears in figure 3.1.

Does this image look like anything? Does it remind you of anything? Drake decided that it does not and would not. And this would imply to

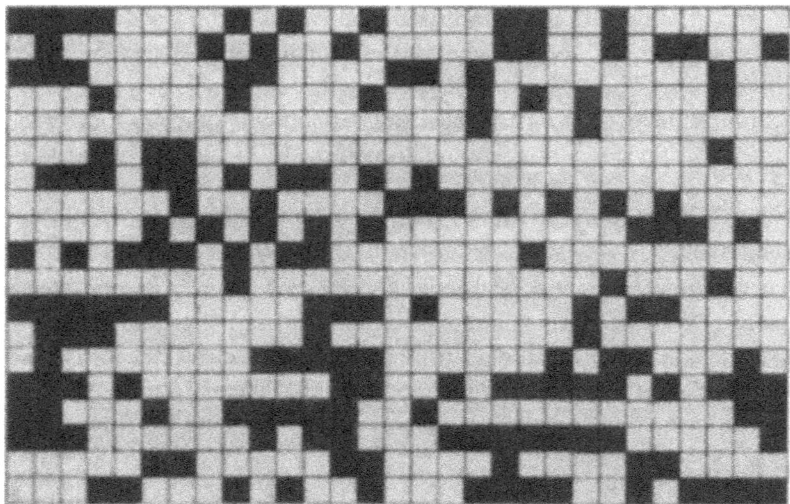

FIG. 3.1 First image of Drake's message. From I. S. Schklovskio and Carl Sagan, *Intelligent Life in the Universe,* translated by Paula Fern (San Francisco: Holden-Day, 1966), figure 30–2, p. 425. Original source: Frank Drake, *Is Anyone Out There?* (New York: Delacorte Press, 1992).

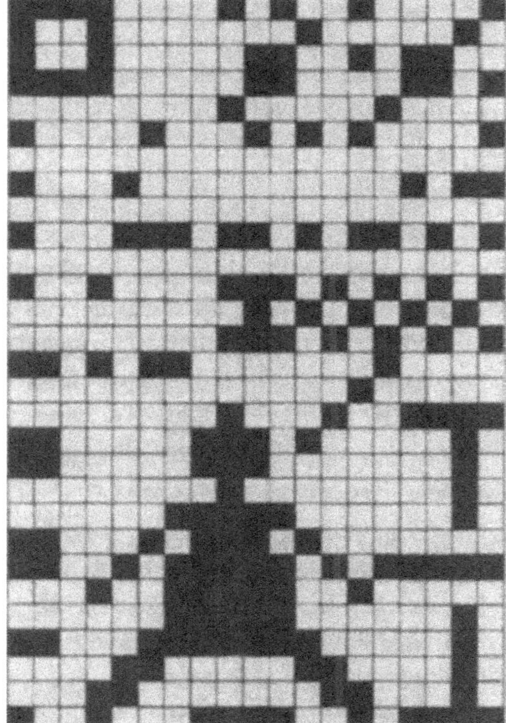

FIG. 3.2 Second image of Drake's message. From I. S. Schklovskic and Carl Sagan, *Intelligent Life in the Universe,* translated by Paula Fern (San Francisco: Holden-Day, 1966), figure 30–3, p. 425.

the receiving civilization, he thought, that it had taken a wrong turn. What if we reverse dimensions, so that the numbers are arranged into rows of 19 across? The result is shown in figure 3.2.

Do you see the humanlike figure at the bottom of the image?[59] For Drake, who concocted the thought experiment in the first place, the image in figure 3.2 proved that two completely "alien" civilizations, spread across the galaxy, might indeed find a way to communicate. Sagan, who presents and comments upon Drake's experiment in his book, agrees. There are many reasons to be skeptical, however. The design of the experiment relies on *so many* assumptions about how the receiving civilization would, or would have to, handle the signal. For example, the assumption that extraterrestrial beings will have eyes is dependent, in my view, on a too restrictive notion of natural selection. Then there is the presupposition that alien beings would *want* to communicate with us, and that they would understand our strategies of decoding messages—strategies, to reiterate, that not even Drake's closest friends, who shared his cultural and scientific assumptions, thought to use. And what if an intelligent creature managed to decipher the above message, but the bodies of those beings were wormlike, with ten eyes and a more complex, non-bilateral structure? Would such a creature be able to interpret the humanlike creature in the image as a bodily form? (There is nothing in Sagan's view, after all, to rule out the possibility of different body plans.) Beyond all these big questions, one doubts that any civilization would be able to correctly guess the many steps required, from an emphasis on prime numbers, to transforming semiprimes into a two-dimensional diagram, and so on.

In short, Sagan's (and Drake's) work functioned at a highly speculative level with unclear consequences. The period of the 1960s and '70s was a contradictory one in the extraterrestrial life debate. On the one hand, this period witnessed unprecedented space exploration, most notably the first Moon landing in July of 1969. Three years later, the Apollo 17 spacecraft captured what came to be the most famous image of Earth from space—the so-called "Blue Marble." The frenzy of space-related activity during this period was marked by (one could even say marred by) the geopolitical climate of the time. In the context of the Space Race, NASA was to a large extent a Cold War project, and part of that story consists of the imperial ambitions of the United States and the Soviet Union. When the Blue Marble photograph was taken, Africa was pictured at the top, but the image was inverted before circulation. Meanwhile, the United States built the world's largest single-aperture telescope, as well as a radio telescope, in the "unincorporated territory" of Puerto Rico, a colonized series of islands that has long served as a laboratory of the United States. The flurry of science-fiction books (*Solaris* [1961], *Dune* [1965], *The Left Hand of Darkness* [1969]), films (*Alphaville* [1965], *2001: A Space Odyssey* [1968], *Planet of the Apes* [1968]), and television shows (*Star Trek* first aired in 1966) during this period frequently portrayed a universe of warring factions, deception, attempted treaties, and violence. And it was during this period, too, that figures like Sagan rose to prominence, becoming household names.

On the other hand, little actual searching for extraterrestrials took place during the same time frame. At least in the United States, the available resources dedicated to "fringe" sciences were totally consumed by the wave of UFO sightings—this is a fascinating topic, which I cannot cover here. At the same time, hard-nosed researchers had begun to develop ways of thinking about extraterrestrial life that were more palatable to the scientific establishment. The historian Steven J. Dick argues that the Space Age brought together biology and astronomy in an unprecedented manner. Based on several factors, including new understandings of DNA and the discovery of amino acids and complex organic molecules in space, biology strove toward universal applicability.[60]

Looking back from the present, one can see the way that the various trajectories, first laid out during the Space Age, have continued to overlap or bifurcate. In the last twenty years, astrobiology has established itself as a respectable academic discipline, with peer-review journals and university courses. Astrobiology benefited greatly from the confirmation, in the 1990s, of exoplanets, that is, planets revolving around stars in other solar systems. That discovery, which essentially means that our solar system is replicated many times in the universe, has been interpreted by many as a modern version of the plurality of worlds. However, although some scientists are optimistic about the sheer magnitude of Earthlike planets, others emphasize the less encouraging finding of exoplanet research: namely, that the vast majority of such planets are only Earthlike in the most abstract sense, and those astronomical bodies have a "diverse range of often-bizarre environments."[61] According to Sara Seager, for example, "It seems that less than 10–20 percent of Sun-like stars could host solar system copies. Instead, astronomers have found that exoplanets and exoplanetary systems are incredibly varied, with planets of nearly all conceivable masses and sizes as well as orbital separations from their host star."[62]

A different kind of question concerns the value of searching for Earthlike planets in the first place. Might it not be possible, after all, for life to exist in environments dissimilar to our own? The origin of life remains radically mysterious, and even the most erudite commentators on this topic acknowledge serious lacunae in our understanding. Most scientists believe that life requires, at the very least, an interaction of organic compounds in water. Current thinking on the topic holds atomic carbon to be the likely building block of life anywhere in the universe, since it is the element with the most potential to form complex molecules—for example, DNA. That said, many remain skeptical about generalizing a definition of life from a single planetary ecosystem. As the anthropologist Stefan Helmreich observes, the language employed by NASA is saturated with uncertainties. He quotes from their astrobiology website: "On Earth, all forms of life to need water to survive. It is likely, *though not certain*, that if life ever involved on Mars, it did so in the presence of a long-standing supply of water."[63] There's a lot we do not know.

The history of the extraterrestrial life debate in the twentieth century leaves us with the impression that the hard science of astrobiology has replaced the wildly speculative endeavors of SETI. And while there is some truth to this story, there are also several reasons to doubt it. For one thing, astrobiology remains a speculative discipline off-center from mainstream scientific inquiry. To offer only the most famous example, there is no consensus about the existence of life on Mars to this day. The gap between astrobiology and SETI is also often overstated, as is evident from even a cursory look at recent articles in one of astrobiology's two major academic journals. To provide just one example, a 2021 article in the *International Journal of Astrobiology* (published by Cambridge University Press) advocates greater attention to the search for intelligent, yet pre-industrial societies, hypothesizing that life forms falling midway between microbes and the kind of highly advanced civilizations capable of sending coded messages may be quite common. The authors propose observing exoplanets for practices such as seasonal agriculture and urban fires, among other possibilities.[64]

For our purposes, the most important problem with the historical account of astrobiology replacing SETI is that SETI has not gone away—far from it. On the contrary, the twenty-first century has witnessed a startling resurgence of SETI projects. In 2015, entrepreneur Yuri Milner donated $100 million towards a new SETI project he calls Breakthrough Listen (notice the project's name). The issue, then, is not that SETI disappeared, but rather that it continues to struggle for legitimacy. A statement of the project, released at the time of its official launching, opens as follows:

> Who are we?
>
> A mature civilization, like a mature individual, must ask itself this question. Is humanity defined by its divisions, its problems, its passing needs and trends? Or do we have a shared face, turned outward to the Universe?

After a few paragraphs of description, the statement concludes, "The 20th century will be remembered for our travels within the solar system. With cooperation and commitment, the present century will be the time when we graduate to the galactic scale, seek other forms of life, and so know more deeply who we are."[65]

The statement was signed by an impressive, if motley, group of individuals, including three Nobel Laureates (in scientific disciplines), Stephen Hawking, Frank Drake, the current director of the SETI Institute (Jill Tarter), the current world chess champion (Magnus Carlsen), two astronauts, one cosmonaut, an actor, and a soprano. Much of the language during the launch centered on the imperatives of what a mature society *ought to do*. At the public launch, for example, Hawking declared, "In an infinite universe, there must be other occurrences of life. Somewhere in the cosmos, perhaps, intelligent life may be watching. Either way, there is no bigger question. It's time to commit to finding the answer, to search for life beyond Earth. We must know."[66] The urgency of the question is palpable. Where Kant saw humanity's insolubility as an aporia in thought, Hawking activated the world's intelligence resources with a kind of "now or never" gambit. And yet, like everyone else before him, he died without knowing. Will the future of humanity be any different?

Impossible Exchange: Colonization, Politics, and Philosophy

As I was writing this chapter, one key member of the Project Breakthrough team published a book claiming that scientists had found evidence of the first interstellar object in 2017, and that what they detected was a bit of discarded alien technology (probably a "lightsail," an object propelled by light). The book's claims have been widely refuted, and what the reader notices perhaps above all else when reading Avi Loeb's *Extraterrestrial: The First Sign of Intelligent Life Beyond Earth,* is its colonial framework. After bragging about his former work with the Israeli defense force and the US military under Reagan, Loeb proclaims that human colonies on other planets will probably look like little Israeli settlements in Palestine. Only time will tell whether Loeb and Milner will be the first humans to make contact with extraterrestrials. But based on the colonial assumptions of *Extraterrestrial,* we may certainly hope not.

Loeb is obviously not alone in his colonial commitments. As this chapter has shown, the extraterrestrial life debate has continually adopted colonial rhetoric for cosmic purposes. The first section of Robin Hanson's seminal essay on the Great Filter is titled "Life Will Colonize." He means this literally, as a statement about what life will ultimately and inevitably do. "All known life seems to have a 'dispersal phase' to encourage colonization," he writes. A great deal of thinking about extraterrestrials has directly shifted earthly imperialist discourse into space. Many have gone further, assuming that extraterrestrials inevitably mirror humans in basic, if narrow, ways: aliens are assumed to have recognizable bodies and sense-organs, as well as recognizable thought processes and desires. Chua and Rehding remark about Carl Sagan that he "optimistically assumed that alien civilizations must be technologically superior to us—that they can do everything humans can do, and then some, . . . then even more. Sagan's aliens should have no problem with the Golden Record. They are 'human+' beings."[67]

A parallel trajectory, however, has insisted upon the radical unknowability of extraterrestrial life. Since Kant, thinkers on this side of the divide have emphasized a single point: we know only life on Earth, and generalizing from a single case is never wise.

A new, third position attempts to overcome the dilemma of the second without the necessity of discovering life on another planet. By *creating* life, scientists such as Kerstin Göpfrich employ techniques from synthetic biology not merely to copy life, but to engineer cells from the bottom up, thereby discovering "completely new ways of assembly, information propagation and replication." With this approach, which has come to be known as "DNA origami," scientists claim the ability to "probe boundary conditions of life."[68]

The realization that humans may soon be capable of producing life has led, in many instances, to profound and unanticipated anxieties. If life can be created, so one argument goes, then perhaps we were also created—not by God or an Almighty Being, but perhaps by some (very) intelligent species. A related anxiety is that if we ourselves may be creations of intelligent beings, then what we perceive as reality may have been created by those same beings. If it strikes the reader as preposterous, the claim is nonetheless important, if only by virtue of its wide circulation and leveraging by influential figures. For example, in a series of widely publicized comments, Elon Musk argued that we are probably living in a simulation. Based on what we know about the current status of human technological innovation, he surmises that "if you assume any rate of improvement at all, then games will be indistinguishable from reality, or civilization will end. One of those two things will occur. Therefore, we are most likely in a simulation, because we exist."[69] As many have commented (in the popular press and elsewhere), Musk's is a version of an argument by the prominent Oxford philosopher, Nick Bostrom, who in 2003 wrote an article explicitly related to the Great Filter theory. In that article, Bostrom suggested that one of the following three hypotheses is necessarily true:

(1) The human species is very likely to go extinct before reaching a "posthuman" stage.
(2) Any posthuman civilization is extremely unlikely to run a significant number of simulations of their evolutionary history (or variations thereof).
(3) We are almost certainly living in a computer simulation.[70]

Bostrom's provocation is that, if we accept that in the future we will probably become "posthumans" who decide to run computer simulations of our own history, then we are most likely those simulations. From this,

Bostrom argues that we probably are living in a simulation. This has come to be known as the Simulation Argument (SA). The idea of posthumans simulating their (or our) history is a strange one and is based in part on Bostrom's thoughts on the potential of computational power: "We noted that a rough approximation of the computational power of a planetary-mass computer is 10^{42} operations per second, and that assumes only already known nanotechnological designs, which are probably far from optimal. A single such a computer could simulate the entire mental history of humankind (call this an *ancestor-simulation*) by using less than one millionth of its processing power for one second."[71]

Unsurprisingly, Bostrom has been widely criticized on many fronts. The assumption that posthumans would necessarily want to run simulations of their evolutionary history seems like a stretch, and some have responded simply to the indulgence of this kind of analytic philosophy. People are dying, some respond, while Bostrom sits in his office in Oxford wondering whether we are all simulations (to say nothing of Elon Musk).

The physicist Marcelo Gleiser makes precisely this moral argument against Bostrom and Musk, before introducing what he sees as the central flaw of any Simulation Argument: if we are indeed simulations of some posthuman civilization, then what is to say that that posthuman civilization is not itself simulating another?[72] From there, we have an infinite regress, in the same vein as the old theological dilemma: if God made us, then who made God? When deployed by atheists, the argument is that there's no reason to believe in God at all, since saying that "something must have made us" merely defers the question: one cannot logically argue that something must have made us and maintain that God just "is." If God just "is," then why not say that we just "are"? There is a reason that more people believe in God than in the simulation argument, however. People may be willing to continue believing in God even while staring a theological aporia in the face, because they accept God as a first cause. As illogical as it may sound to an atheist, millions of people the world over believe in God's primordiality based on faith. But why would anyone have faith in a posthuman civilization? It's difficult to imagine how anyone would make a first cause argument about a super intelligent civilization in secular terms. In the remainder of this chapter, I follow the two threads from the preceding discussion: first, the accusation against some philosophers that they ignore the problems of the world; second, the charge of infinite regress.

In relation to the first thread, there is a twist. In my own view, criticisms of Bostrom would not be so damning (they seem fairly petty) were he simply speculating in the ivory tower and writing the occasional journal article. That is not the case, however. Bostrom is director of the Future of Humanity Institute, an organization that has immense real-world power. Most importantly, the Future of Humanity Institute creates guidelines for how and where the superrich should donate their money. (This chapter was written before the scandal surrounding Samuel Bankman-Fried and, therefore, mercifully ignores that segment of the story.)

At the 2015 conference, Effective Altruism Global, existential risk rather than current problems was the main topic of conversation. That thematic emphasis is attributable, in large part, to Bostrom's work, and not only to his Simulation Argument. In a different paper, Bostrom coined a term that has become fundamental to the Effective Altruism movement: existential risks (or "x-risks"). By Bostrom's definition, x-risks are of a different order from all other risks, including genocide (which Bostrom calls a "local risk") and the thinning of the ozone layer (which he calls a "global risk"). An x-risk is: "One where an adverse outcome would either annihilate Earth-originating intelligent life or permanently and drastically curtail its potential."[73] Because x-risks are by definition permanent and irreversible, they are to be avoided at all costs. As Amia Srinivasan notes, followers of Bostrom, including many of the Effective Altruism Global attendees, believe that it is "better to try to prevent some hypothetical x-risk, even with an extremely remote chance of being able to do so, than to help actual living people."[74] Srinivasan continues:

> X-risks could take many forms—a meteor crash, catastrophic global warming, plague—but the one that effective altruists like to worry about most is the "intelligence explosion": artificial intelligence taking over the world and destroying humanity. Their favoured solution is to invest more money in AI research. Thus the humanitarian logic of effective altruism leads to the conclusion that more money needs to be spent on computers: why invest in anti-malarial nets when there's a robot apocalypse to halt? It's no surprise that effective altruism is popular in Silicon Valley: PayPal founder Peter Thiel, Skype developer Jaan Tallinn and Tesla CEO Elon Musk are all major financial supporters of x-risk research.[75]

Theorizing x-risks and the robot apocalypse is no mere abstract indulgence. On the contrary, the intellectual labor expended on such topics has tremendous real-world influence.

This relates to Achille Mbembe's observation that "abstract theory has never had such a hold on the material and social reality of the world as it does today." "Theory," says Mbembe, "is always a particular theory of the world. And that world is being constructed more and more by entities such as finance capital and abstract singularities like derivatives."[76] To Mbembe's examples of finance capital and derivatives, we can more prosaically add the direct deployment of philosophy by tech entrepreneurs.

In closing the second loop (the one that seems always to result in an infinite regress), it is helpful to note the resemblance between twenty-first-century simulation arguments and older hypotheses. As many have commented, Bostrom's 2003 paper has a great many philosophical precedents, ranging from Zhuangzi's "dream of the butterfly" to Descartes's radical doubt. But the more relevant predecessor, I want to suggest, is something that journalists picked up on, especially when Musk began spouting his version of the argument. I am talking about the Matrix. (This much is certain: "simulation hypotheses" are not the consequence of twenty-first-century technologies from Silicon Valley.)

In 1999, on the verge of a new century and during a time of tremendous anxiety (Y2K, millenarian cults, and so on), the film *The Matrix* captured the time perhaps more than any other cultural product. The film is about a group of hackers who discover that humanity is living inside a simulated reality. That simulated reality, the hackers learn, was created by a civilization of hyper-intelligent machines who are nourished by human bodies. The film follows the group of freedom fighters as they take on the system and attempt to liberate themselves and humanity from the Matrix.

After the film's release, many were quick to compare *The Matrix* with the work of one of the most famous philosophers of the 1990s, Jean Baudrillard. Indeed, the film's directors (the Wachowskis) made the connection explicit within the film itself: in one scene, the protagonist Neo (played by Keanu Reeves) uses a copy of Baudrillard's *Simulation and Simulacra* to hide illicit items. In the popular imagination, *The Matrix* and Baudrillard's theories became synonymous.

Published in French in 1981 and translated into English two years later, *Simulacra and Simulation* marked a high point of French theory in the anglophone world. Along with the works of Derrida, Foucault, Lyotard, Deleuze, Cixous, and others, between the late 1980s and the early 2000s, Baudrillard's work had a tremendous impact not only on American academia, but also on filmmakers (as the example of *The Matrix* shows), music, and art. Linking Baudrillard's writings to *The Matrix* also happened in conjunction with two globally transformative events: the end of the Cold War (which resulted in the establishment of US global dominance), and the rise of the internet (along with the dot-com generation).

Those of us who came of age in the 1990s had already experienced a Simulation Argument–like shock. Along with the internet's ubiquity and the hyper-mediatization of global events (including war), *The Matrix* compelled viewers to face the prospect that—strange as it may seem—we are living inside a simulation.

But those who were quick to draw an analogy between *The Matrix* and Baudrillard's writings failed to notice something important: their fundamental contradiction. As Baudrillard himself said in a 2004 interview,

> Certainly there have been misinterpretations, which is why I have been hesitant until now to speak about *The Matrix*. The staff of the Wachowski[s] contacted me at various times following the release of the first episode in order to get me involved with the following ones, but this wasn't really conceivable (laughter). Basically, a similar misunderstanding occurred in the 1980s when New York-based Simulationist artists contacted me. They took the hypothesis of the virtual for an irrefutable fact and transformed it into a visible phantasm. But it is precisely that we can no longer employ categories of the real in order to discuss the characteristics of the virtual.[77]

The important part comes in the last sentence: we can no longer employ categories of the real to discuss the characteristics of the virtual. What Baudrillard means by this (and he emphasizes it throughout *Simulacra and Simulation*), is that the world cannot be grasped as a bifurcation between the virtual and the real. And *The Matrix*, by contrast, insists on precisely this bifurcation: Neo and his fellow freedom fighters realize that they are in the Matrix, and they struggle against it to reenter the real world. Baudrillard, for his part, believed that we had entered a single plane of reality, where bodies, images, and simulations coexist—no real, no virtual, only a world of simulacra. (A simulacrum is often defined paradoxically as a copy without an original.)

Musk and Bostrom also take an anti-Baudrillardian stance when they suggest that we are probably simulations but that some posthuman civilization is not. For Baudrillard, by contrast, it was simulation all the way down. What Baudrillard wanted so urgently to show is our total disorientation in a world where the boundary between reality and virtuality has collapsed—totally, without any super-intelligent civilization behind the scenes keeping things going. Because we no longer have a space of reality to rely on (even, that is, conceptually), we must learn to navigate a world of free-floating simulacra.

Philosophically, *The Matrix* and *Simulacra and Simulation* could not be more different. And there is another reason that the two do not pair well together: the book (the French original) was published eighteen years before the film appeared. In some sense, then, *Simulacra and Simulation* and *The Matrix* participated in different historical moments. *The Matrix* was a quintessentially post–Cold War, turn-of-the-century product. *Simulacres et Simulation* was published during the late Cold War period (in 1981) and is better understood in relation to Baudrillard's preceding work on post-Marxism and commodity culture.

But *Simulacra and Simulation* was not Baudrillard's last word on the topic. In 1999—the same year that *The Matrix* was released—Baudrillard published a fairly obscure little book called *Impossible Exchange*, which was to become my favorite. In it, he writes,

> Everything starts from impossible exchange. The uncertainty of the world lies in the fact that it has no equivalent anywhere; it cannot be exchanged for anything. The uncertainty of thought lies in the fact that it cannot be exchanged either for truth or for reality. Is it thought which tips the world over into uncertainty, or the other way around? This in itself is part of the uncertainty.
>
> There is no equivalent of the world. That might even be said to be its definition—or lack of it. No equivalent, no double, no representation, no mirror. Any mirror whatsoever would still be part of the world. There is not enough room both for the world and for its double. So there can be no verifying the world. That is, indeed, why "reality" is an imposture. Being without possible verification, the world is a fundamental illusion.[78]

The argument of *Impossible Exchange* is obviously related to Baudrillard's earlier thinking about simulacra. But the figure of exchange (and non-exchangeability) adds a layer of complexity. In some sense, it would seem that Baudrillard is making a point similar to Kant's vis-à-vis insolubility: because we do not know any other worlds (or because we have no knowledge of other rational beings), nothing we say can be verified. On the other hand, Baudrillard seems to up the ante on Kant. For Kant, knowledge of another rational being is the condition that would make the human character soluble. For Baudrillard, by contrast, discovering a rational being on another planet would not change much, because that being would be part of our knowable reality and, therefore, our "world." For Baudrillard, in other words, even an alien being would not provide a viable basis for comparison. "Any mirror whatsoever would still be part of our world," he writes, implying, I think, that any *döppelganger*-like alien being would be, too.

According to a certain strain of classical philosophy, the search for new knowledge is described as a kind of dead-end street. Meno's paradox, for example, holds that if you know what you are looking for there's no need to look for it, and if you don't know what you are looking for there's no point in looking—new knowledge is therefore impossible. Plato's *Meno* (written as a dialogue between Meno and Socrates) presents no way out of the dilemma. In concluding, the best that Socrates can offer is a plug for bravery: "We will be better men, braver and less idle" if we keep on searching, he says, even if we never get anywhere.[79] In the end, Socrates admits that new knowledge is fundamentally elusive. If anything, Meno's paradox is ratcheted up a few notches.[80]

Accepting the truth of Meno's paradox means accepting solipsism: being stuck in a world of one's own without any exit route or any possibility of seeing beyond one's own field of vision. And in modern philosophy, too, the paradox is often characterized as a vicious circle, a cruel hall of mirrors, a suffocating chokehold.[81]

Baudrillard's notion of impossible exchange moves in a different direction. For him, the fact that every thought takes place on a single plane with nothing preceding it or beyond it leads not to solipsism, not to suffocation, but to disorientation and enigma. Reality is an imposture, he says, and the world a fundamental illusion. From a Baudrillardian perspective, neither resignation nor bravery are appropriate. Rather, he proposes three basic theorems:

- The world was given to us as something enigmatic and unintelligible, and the task of thought is to make it, if possible, even more enigmatic and unintelligible.
- Since the world is evolving toward a frenzied state of affairs, we have to take a frenzied view of it.
- The player must never be bigger than the game itself, nor the theorist bigger than theory, nor theory bigger than the world itself.[82]

I do not mean to say that these theorems somehow solve a two-millennia-old paradox. And, indeed, the reader may be surprised by how much real estate I've given Baudrillard in this chapter, especially considering that he has fallen out of fashion. Many who consider themselves "serious" intellectuals find no value in Baudrillard's later work (often demarcated by anything after *For a Critique of the Political Economy of the Sign*): at a conference, I once heard a famous theorist say that in the 1980s Baudrillard's work "degenerated into I-don't-know-what." And that was a statement by a critical, Leftist thinker. Even less generously, Baudrillard's entire oeuvre was famously dismissed as a prime case of "fashionable nonsense."

I'm aware of how Baudrillard's work can come off, but I remain compelled by it, if only because it energized so many writers, artists, and filmmakers (even if they completely misinterpret his work, they are still inspired!). Given the ultra-violent history of space exploration, a history that is nearly equivalent to the history of colonialism itself, Baudrillard's work offers something different. Instead of omniscience, he advocates affirming enigma. Instead of control, he asks us to accept a frenzied state of affairs. Finally, he asks us to calibrate theory relative to the task at hand. As an elliptical end to this chapter, the three theorems are intriguing. As the beginning of something new, they leave us little to work with—little, but (at least in my view) not nothing.

Final Remarks on Cosmic Loneliness

This chapter has shown that with every move outward into the cosmos, humans have paradoxically become increasingly isolated. In the seventeenth century, Fontenelle dragged and dropped colonial stereotypes in his speculations on extraterrestrials. He could not, in other words, see or hear beyond his own colonial limitations. Despite tremendous strides in astronomy and biology, twentieth-century thinkers found themselves progressively atomized. In one instance, John Lilly put himself in an isolation tank in order to explore the beyond—but this time it was the beyond within. Many others in the twentieth century continued to espouse colonial logics in ways I have documented in these pages. Moving into the twenty-first century, things seem only to have deteriorated further, to the point where researchers openly declare their commitment to extraterrestrial settler colonies, and where private funders of space research doubt the existence not only of outer space, but of physical reality itself. Baudrillard's theories push solipsism to its tipping point—perhaps starting to break it open from within. I personally find his answer to the dilemmas of knowledge (encapsulated in the three theorems) intriguing, but it's difficult to see how he might help us. He provides no roadmap for how to use or elaborate his ideas.

In *The Human Condition*, Hannah Arendt lamented the fate of "modern man." "When he lost the certainty of a world to come," she wrote, modern man "was thrown back upon himself and not upon this world; far from believing that the world might be potentially immortal, he was not even sure that it was real. . . . What men now have in common is not the world but the structure of their minds, and this they cannot have in common, strictly speaking; their faculty of reasoning can only happen to be the same in everybody."[83] This chapter has shown that the crisis of "modern man" has only deepened and intensified over the past two hundred years. Two important caveats remain, however.

First, I do not mean to cast doubt on the scientific enterprise *in toto*.[84] Careful, empirically based research is more important now than ever. But researchers working in the "hard" sciences need also to be more aware of their hidden assumptions, and that is where a book like this one seeks to intervene.[85]

Second, if this chapter has addressed humanity's cosmic isolation, it is only a certain "genre" of the human who feels this way.[86] Many people experience love and intimacy with animals, even in the absence of symbolic language. And not all people everywhere feel a disconnect from the universe in the way some modern astronomers do.

I do not believe that in attempting to speak about the alien we are just speaking about ourselves. What seems clear from the material presented in this chapter, though, is that any genuine attempt to speak about something beyond who we are requires a deep, thorough, and hard look at who we think we are. Without that critical step, we are doomed to a life of ceaselessly comparing ourselves to nothing, eternal silence ringing in our ears.

4
An Alien Music

According to experts on the topic, it is only a slight exaggeration to say that music saved the whales—that's how I began chapter 1 of this book. But experts are not always right. In that first chapter, I largely assumed the veracity of the expert claims: that the discovery of whale song led to a reconceptualization of cetaceans and, ultimately, to an international prohibition on whaling. Black feminist author Alexis Pauline Gumbs has her doubts, however. In, *Dub: Finding Ceremony,* a book that falls somewhere between poetry and nonfiction prose, she writes,

> between you and me, we knew it would never work. just because the singing of the whales had caused bumper stickers and rallies and international bans on their murder and the criminalization of the exploding harpoon (you know. that thing that got under their skin and destroyed them from the inside) didn't mean it would work for us. i mean how long had we, black women, been singing.
>
> when they decided the whale was an intelligent creature, nuanced, descriptive, they decided that the people who killed them were greedy, were savage, were less evolved. isn't that interesting. the same people who forced the whaling indigenous into sale instead of ceremony now spoke of evolution. spoke of the humane and didn't choke. this is why we didn't have much hope. our intelligence and the multiple forms of proof required did not inspire the world to disentangle its hooks from our looks and our attitude.
>
> we assert that it was not the song of the whales that saved them. if singing could save we'd be god. it was the fact of other sources of oil to move into, other deep black resources to extract. it was a fact. they could only save the whales once they knew they didn't need them. it was a simple as that. and they haven't found a way yet to say it. their needles in our skin, targeting us where we breathe. which is everyone we love. trapping

us below and yet detracting us above. chasing us across oceans. they risk their very souls. they know it though. they need us more than gold.[1]

To be sure, the history of whaling and anti-whaling is complex, many-sided, and nonlinear: it would probably be a mistake to attribute the end of whaling *either* to the discovery of whale song *or* to the "fact of other sources of oil to move into." What is more, whaling has not ended: Norway and Japan still engage in whaling activity, and several species of whales remain endangered to this day. But Gumbs makes two incisive interventions. First, it is striking how easily some have accepted the "whale song saved the whales" narrative without examining shifts in production over the course of the twentieth century that made whale oil redundant—as though discussions of activism and legal proceedings can be untethered from economics. Second, she points to a contradiction: if discovering whale song led to greater value placed on whales' lives, then why have Black women suffered a different fate? If the value of Black life is not equivalent to the valorization of Black music, then why should the case be any different for whales?

Perhaps it goes without saying that human discussions about animals and aliens say more about humans than about either animals or aliens. We are a perplexing bunch—especially to ourselves. And because we know of no other rational being, said Kant, we have nothing to compare ourselves to and therefore can never define who or what we are. But if Kant pointed to our inability to define ourselves through positive comparison, he omitted the other side of the story: the way the human (or, perhaps better, the Human) has continually defined itself through negative comparison, through comparison with a negative term. Until the early modern period, that negative term was typically the animal. European Christian thought positioned the human as lying midway between God (at the highest point) and beasts (at the lowest). During the early modern period, however, colonial subjects and slaves came increasingly to stand in for the negative image against which rational Europeans measured themselves. Ronald Judy opens his devastating critique of European humanism with two quotes from Kant.[2] Here is the first:

> The Negroes of Africa have by nature no feeling that rises above the trifling (*Die Negers von Afrika haben von der Natur kein Gefühl, welches über das Läppische stiege*). Mr. Hume challenges anyone to cite a single example in which a Negro has shown talents and asserts that among the hundreds of thousands of blacks who are transported elsewhere from their countries, although many of them have even been set free, still not a

> single one was ever found who presented anything great in art or science or any other praiseworthy quality, even though among the whites some continually rise aloft from the lowest rabble, and through superior gifts earn respect in the world. So fundamental is the difference between these two races of man, and it appears to be as great in regard to mental capacities as in color.[3]

In the second instance, Kant comments on a statement made by a "Negro carpenter" as reported by one Father Labat. "This fellow was very black from head to foot," says Kant, "a clear proof that what he said was stupid."[4] As Judy remarks, Kant's arguments—To be Negro is to be stupid; To be Black is to be Negro—are not deduced from sensible grounds and would be considered illegitimate on the terms of Kant's own philosophy. That said, in his rigorous, seventy-page exegesis, Judy shows that Kant's arguments about "the Negro" are not mere sloppy thinking: they show something fundamental about the architectonics of Kant's critical philosophy.[5]

Kant's statements about Blackness are important because they illustrate the way that rational, European Man has constructed itself in opposition to the *a priori* stupidity of the Negro. The human is defined implicitly as that being who *does* possess a "feeling that rises above the trifling." (*Läppische* can also mean worthlessness, childishness, or silliness.) The "Negro," by contrast, is that paradoxical being who is human but who nonetheless fails to do anything beyond the trifling. In this sense, the Negro is a human who is no human at all.

In the twentieth and twenty-first centuries, it is the White, Western, rational Man who either listens to the cosmos or sends something into it in an effort to make contact with an alien being. Think of all those dozens of sci-fi films and novels where an alien civilization must go through a White American (a president, a military commander, or a linguist) as a first point of entry to humanity. And recall John Lilly's musings about making contact with alien intelligence—by which he meant any nonhuman, whether extraterrestrial or marine. Once communication with dolphins had been established, he wrote, they will for a long time "be in the position of the Negro races in Africa who are attempting to become Westernized," and who must "prove their usefulness in those things which the human races in general attempt to achieve."[6] Amazingly, Lilly never once said that extraterrestrials would need to prove their usefulness to humanity. The reason for this is simple: while he was certain of the inferiority of the "Negro races," he equally assumed the superiority of extraterrestrial beings. To borrow Chua's and Rehding's comments on Sagan,

Lilly assumed that alien civilizations must be technologically superior to us—"that they can do everything humans can do, and then some."[7] In his schema, the human is defined by its place on an evolutionary axis, where the extraterrestrial is the human+ and the Negro is the human−. (The dolphin, to recall chapter 1, has equivalent intelligence to humans but took a different evolutionary path.)

These historical observations help situate the essay by Ronald Judy to which I've referred. Judy begins his essay with a quote not from Kant but from the Black Martinican philosopher Frantz Fanon, the first line of which reads, "At the risk of arousing the resentment of my brothers of color, I shall say that the Black is not a human."[8] Judy remarks, "Today the proclamation that the black is not a human is not guaranteed to arouse resentment, even among 'brothers of color,' the way it might have thirty-nine years ago when Fanon wrote these lines. This is due in some measure to the generally accepted perception of the Black as essentially a victim of humanism. And that perception rests upon a particularly arrogant comfort about the ends of humanism: the inevitability of the effects of humanity."[9]

Taking Fanon and Judy as a hyperbolic starting point, this chapter has two primary motivations. The first is largely theoretical and pivots on a single question: What does the Fanonian critique picked up by Judy mean for the question of interspecies communication? How do we even begin to think about human attempts to communicate with other beings when the notion of the human has been dealt such a devastating blow? Recalling Gumbs's intervention, we might ask how to think meaningfully about human and animal capacities when, for some, it seems easier to declare the value of whales' lives than to say "Black lives matter." The second motivation is more historical. As the Order of the Dolphin (the group led by Drake and Sagan) was meeting to launch the first formal SETI project, African Americans were struggling for basic civil rights, and African countries were clamoring for independence after centuries of colonial rule. At the moment that Lilly was using NASA funding to pay poor, Black Caribbeans to build the Communication Research Institute, White American political scientists were questioning whether Africans were capable of self-rule, and the CIA was assassinating African leaders like Patrice Lumumba to make sure that they could not be. This too-brief chapter provides a glimpse into the other side of the SETI debate—not by speculating about alien or dolphin intelligence, but by listening to responses by members of the human species whose humanity has been derided by the same scientific community that has continually attempted to speak with animals and aliens.

Make no mistake, it is not simply a matter of illustrating the simple-mindedness and arrogance of racist Whites, as though the problem of the Human that Fanon and Judy diagnose is a matter of becoming more liberal. As we have seen throughout this book, the discourse of interspecies communication is saturated with racist thinking; one might say that race is constitutive of interspecies communication theory. As one author famously put it, the "problem of the Negro is a problem for thought."[10] Jared Sexton goes even further, asking after European humanism's relation to Blackness in a way that is both wonderfully elliptical and precise. "What is the nature of a form of being that presents a problem for the thought of being itself?" Sexton asks. "More precisely, what is the nature of a human being whose human being is put into question radically and by definition, a human being whose being human raises the question of being human at all?" After a few sentences, he continues, "How might it be thought that there exists a being about which the question of its particular being is the condition of possibility and the condition of impossibility for any thought about being whatsoever? What can be said about such a being, and how, if at stake in the question is the very possibility of human being and perhaps even possibility as such? What is the being of a problem?"[11]

I approach these questions by considering Black musicians and artists who, I believe, have offered a potent rejoinder to decades, if not centuries, of (White) humanism. I have in mind creative practitioners falling into a category sometimes referred to as Afrofuturism, although I use the term loosely and primarily as a heuristic. Afrofuturism is usually understood as a genre of Black creative practice that attempts to imagine a world beyond this one—an alien world, that is. The term *Afrofuturism* is typically attributed to Mark Dery (and to an article of his from 1994), but Dery was only reflecting on decades of Black music, literature, and art that had explored alien and futuristic themes: Sun Ra's intergalactic research, Jimi Hendrix's "Astro Man," Octavia Butler's Xenogenesis series, and Anthony Braxton's *Trillium R*, to name just a few examples.

This chapter presents one example from the Cold War Space Race period, and another from the more recent period of twenty-first-century tech entrepreneurialism. The first example is Sun Ra, a musician and thinker who continues to be widely influential and who has generated a large body of theoretical literature. The second example is the Canadian artist Kapwani Kiwanga, a more recent figure with whom I have worked sporadically over the past few years. Sun Ra and Kapwani Kiwanga respectively offer counter-histories to Cold War scientists (like John Lilly and Carl Sagan), and to twenty-first-century figures (scientists like Avi Loeb, entrepreneurs like Elon Musk, and philosophers like Nick Bostrom). Unlike the White, male scientists who dominate the field and focus either on linguistic or mathematical models, the Black thinkers in this chapter turn most frequently to music in their philosophical-political interventions.

Consider, for example, Kodwo Eshun's distinction between Soulful and Postsoul music in his seminal text, *More Brilliant Than the Sun*:

> Like Brussels sprouts, humanism is good for you, nourishing, nurturing, soulwarming. . . . Present-day R&B is a perpetual fight for human status, a yearning for human rights, a struggle for inclusion within the human species. [By contrast] the music of Alice Coltrane and Sun Ra, of Underground Resistance and George Russell, of Tricky and Martina, comes from the Outer Side. It alienates itself from the human; it arrives from the future. Alien Music is a synthetic recombinator, an applied art technology for amplifying the rates of becoming alien. . . . From the outset, this Postsoul Era has been characterized by an extreme indifference towards the human. The human is a pointless and treacherous category.[12]

Writing in a deliberately hyperbolic tone, Eshun observes a red thread of much popular culture: the relationship between certain styles of Black music and extraterrestriality. Eshun would seem to agree with Fanon's and Judy's assessment of the human and of humanism, but he goes a step further: if the human is a pointless and treacherous category, then the musical styles he advocates are Alien Music. Does critiquing humanism necessarily lead to affirming the alien? What would an alien music sound like, if indeed it can sound like anything at all? To be sure, what Eshun has in mind differs fundamentally from the kind of thought experiments put forward by Sagan and Drake. How do we begin to ask questions about the alien when the meaning of "we" is put radically into question?

I rely on direct quotation more in this chapter than in others. I do this to foreground the different registers of speech and prose style that populate the text, but also to disrupt narrative neatness since, as I've said, a coherent writerly "we" should not be assumed. The chapter uses quotations galore, but it is also about quotation, in a sense—about who speaks and from where. It is about a claim like the one by Eshun quoted above, that the music of Alice Coltrane and Sun Ra "arrives from the future."

Afrofuturism may seem a strange topic for a book on interspecies communication. Neither Sun Ra nor Kapwani Kiwanga provide a direct guide for communication with other species, nor is interspecies communication a major theme in their work. I turn to their work to show just

how much needs to be cleared away before a step in the direction of interspecies communication can happen. As I asked earlier in this chapter: How do we even begin to think about human attempts to communicate with other beings when the Human itself is built on shaky foundations?

Sun Ra

Sun Ra is perhaps the most prominent musician associated with the movement that came to be known as Afrofuturism. Born Herman Poole Blount in Birmingham, Alabama, in 1914, as an adult in Chicago, he insisted on being called Sun Ra and often asserted that he was not from planet Earth. A prolific composer and virtuosic multi-instrumentalist, Sun Ra performed complex, experimental, and futuristic-sounding music for several decades.

Sun Ra's deployment of Space Age iconography, stage costumes, and lyrics is certainly related to philosophical discussions about Blackness and humanism. But there was also a more prosaic and proximate context for his work: the Cold War Space Race of the 1950s through the 1970s. While Lilly, Sagan, Drake, and others were devising projects to search for extraterrestrial intelligence, the struggle for civil rights, Black liberation, and access to basic requirements was being fought on the ground in Birmingham and Chicago—and, for that matter, anywhere else that Black people found themselves. During the famous Memphis sanitation strike of 1968, Black workers held up signs stating "I Am a Man." In the same year, Sun Ra moved to Germantown, Philadelphia, where he established a new base for his ensemble (the Arkestra) and continued developing his musical and philosophical ideas. Less than a year later, Apollo 11 would land on the moon—if it was a "giant step for mankind," Sun Ra must have thought, then mankind certainly did not include him and other oppressed Black people on the South Side of Chicago, the townships of apartheid South Africa, the slums of Jamaica, or the favelas of Brazil.

Like many in my generation, I was introduced to Sun Ra first through his film, *Space Is the Place*. While the extent of his control over the film's production is contested (as is the extent to which it faithfully reflects his ideas), it remains a useful starting point for discussing his work, especially considering its wide popularity. *Space is the Place* begins with Sun Ra walking through a mysterious planet, strange and unfamiliar plants around him.[13] Over a background of a freely improvised, frenetic horn ensemble, Sun Ra begins to hum, before speaking: "The music is different here. The vibrations are different. Not like planet Earth. Planet Earth sounds of guns, anger, frustration. There was no one to talk to up on planet Earth who understands. We set up a colony for Black people here. See what they can do on a planet all of their own without any White people there. They could drink in the beauty of this planet. We'll bring them here through either isotope teleportation, transmolecularization or better still, teleport the whole planet here through music."

The majority of the film, however, takes place on Earth. A brief synopsis: presumed lost since a European tour in 1969, Sun Ra discovers a new planet (shown in the film's opening), which he finds suitable for habitation by Black people. He returns to Earth to pick up (forcibly if need be) Earth's Black population so that he can transport them to his newly discovered planet. The film ends with Sun Ra's spaceship traveling toward his new planet, leaving Earth's destruction in his wake. (The end of the film bears a striking resemblance to Henry Dumas's "Will the Circle Be Unbroken?" a marvelous 1966 short story about a New York club where the avant-garde jazz obliterates White patrons on entrance.[14]) Several additional subplots are introduced in the course of the film: Sun Ra time-travels to the past, where he plays music at a Chicago strip club; he sets up the "Outer Space Employment Agency" to attract Black youth to his mission; he plays cards against an evil character named the Overseer, who helps a group of NASA scientists kidnap and attempt to assassinate him.

The film also develops several concepts key to Sun Ra's philosophy, most notably that musical vibration is a force capable of transforming, even *transporting* people to other dimensions. Also prominent is the mixing of futurist with Egyptian iconography in the film. Especially in his early and middle years, Sun Ra turned incessantly to ancient Egypt as an exemplar of African civilization. As a young man in Chicago, he read voraciously a slew of texts attempting to overcome centuries of Eurocentric bias about the ancient world. Authors like Theodore P. Ford, George Parker Wells, Godfrey Higgins (all of whom Sun Ra read) critiqued the version of "Western civilization" bequeathed to us by so-called classicists—the progression from Greece to Rome and then the (European) Renaissance and "modernity." A number of the authors Sun Ra read also controversially claimed that ancient Egyptians were Black. If the details of the books and pamphlets read by Sun Ra were dubious, they preempted the major upheaval taking place right now in academic discipline of Classics, where scholars are finally beginning to question established dogmas.[15]

What made Sun Ra so singular was that—offstage—he maintained and lived what some assumed was merely a persona: he walked through the streets of New York and Philadelphia wearing outlandish clothes (including his signature cape and futuristic headgear), confounding interlocutors and interviewers by insisting that he was not born but rather had *arrived* on earth. He seldom said more than that about his origins, though he famously declared on a few occasions that he came from Saturn.

Sun Ra's early revelations about Egyptian history compelled him to join the ranks of the most ardent hermeneuts of suspicion. For the remainder of his life, he deployed tactics of reading and exegesis that sought hidden meanings in texts—even, and especially, meanings deliberately hidden by those in power. He focused his energies on the Bible, deploying a range of quasi-cryptographic strategies, including assigning numbers to letters and then subjecting the letters to permutations and calculations. (This numerical technique has an old and venerable tradition going back to the pre-Christian practice of *gematria*—two millennia, that is, before the popularized version in Michael Drosnin's *The Bible Code.*) Sun Ra also developed a keen interest in fringe and more-than-fringe science: Mesmerism, Swedenborgianism, the theosophy of Helena Blavatsky and Rudolf Steiner, the mystical number symbolism of Georgei Ivanovitch Gurdjieff, and the esotericism of P. D. Ouspensky, among others.[16]

More often than not, though, Sun Ra proclaimed a deep admiration and respect for science—although he defined the term idiosyncratically. His biographer, John Szwed, notes that as a child in Birmingham, Sun Ra "plumbed the science and technology section of the public library (*which in those days was also deep with books on psychic phenomena and the occult*)."[17] As an adult, he explored the boundaries and connections between mainstream science and its repressed offshoots. As Szwed observes, for Sun Ra, "'science' was somewhere between or beyond science fiction and science. More than a method of reasoning and a set of laboratory practices, it was also a mystical process, and (as the rappers imply by 'dropping science') a kind of secret or suppressed knowledge which had the power to create new myths, erase old ones, altering our ratio to each other and the rest of the universe."[18]

In Sun Ra's own words, he was not doing "science as we know it, but another kind. I've been looking for a solution which gets back to Egypt, and to the whole universe. I think musicians are on a superior level, but unlike scientists, they haven't been accepted for their abilities."[19] Sun Ra viewed music as a form of research into social and material reality. His short-lived organization, Ihnfinity, Inc., set out to "perform spiritual-cosmic-intergalactic-infinity research works relative to worlds-dimensions-planes in galaxies and universes beyond the present now known used imagination of mankind" (242). Music, he felt, had the capacity to elevate our knowledge as well as our consciousness, helping us to establish a meaningful place in the cosmos.

What, for Sun Ra, was the relationship between science and truth? This is a topic worthy of further discussion. In an age of Right wing, neofascist "post-truth," we should not and cannot casually celebrate the rubbishing of science. But fetishizing science cannot be the alternative. Sun Ra was wary of established truths. He challenged the veracity of taken-for-granted claims, asking what evidence they were based on and why he or anyone else should believe them. But he also knew that even when something is true, that does not mean it is good. The value of a statement is not solely dependent on whether it is true—after all, truths are a dime a dozen.[20] And truth can also be a bad thing, Sun Ra was wont to say, "as when a man says, 'I'll kill you' and he does." "Those who live by reality are slaves of truth," Sun Ra said. "It's a kind of narcotic" (317). To explain the last remark, he used an apt example from his experiences as a Black man in Birmingham, Chicago, New York, and Philadelphia: "When the police find out the truth about you, they say they get the 'dope' on you" (317). Stop-and-frisk, police harassment, and incarceration all trade on truth just as much as lies. When Black people are put away for years for possession, the truth hardly matters—even when it's "true."

Sun Ra was skeptical about every aspect of his life. From an early age, he felt that his birthname (Herman Poole Blount) did not fit. His first name was a reference to the early twentieth-century magician, Black Herman. His middle name, apparently, was given in honor of one his mother's (male) friends. "There was talk," Szwed comments, "but his mother said Mr. Poole just wanted to have a child named after him." "Blount," finally, "was his patronymic, a connection to a father's line of kin" that he felt distant from (4).

It was not only his birth *name* that he doubted—as he reached adulthood, Sun Ra became increasingly skeptical about the notion of birth. (More could be said, in a quasi-psychoanalytic manner, about what this skepticism meant in relation to the women of his life, especially his mother.) It was possibly while reading the *Egyptian Book of the Dead* that he was introduced to a radically different conception of life on Earth: the dead are never truly dead, they are simply asleep. From this revelation, Sun Ra developed an idiosyncratic philosophy in which death and life received equal scorn. Death disturbed him, but so did life, which he viewed as its mirror opposite. At least in one version of his thinking, life is registered as a kind of pathology unique to Earth. What exists beyond life, beyond the time of historicity, is not death but *being*. Unlike life and death, being is eternal, and to come into (or onto) the world is less a matter of birth than of *arrival*: we arrive on the world, Sun Ra believed, and we depart from it. When asked about specific aspects of his life chronology,

Sun Ra responded: "I don't know the years. I wasn't in a time zone."[21] The author William Sites comments that, for Sun Ra, "Time, in any conventional sense, was a trap. Birth, he claimed, was in fact the beginning of death."[22]

In rejecting the concept of death, he recognized that he was asking for the impossible. And indeed, asking for or even demanding the impossible became a motif in his later years. "I'm talking about something so impossible; it can't be true," he said. "But it's the only way the world's gonna survive, this impossible thing" (295). Impossibility was related to myth, or to mythical time beyond and outside history: "Myth speaks of the impossible, of immortality. And since everything that's possible has been tried, we need to try the impossible" (316). Sun Ra was not asking anyone to give up their life, but rather their concept of it, their concept of *having a life*—and, by extension, of having to die. He shouted during one concert: "If you're willing to give up your life for your country, will you give up your death for me?" (258).

Sun Ra refused the title of a human born in this world. At times, he spoke about being not-of-this-world in a more colloquial sense, as is common in descriptions of unusually talented people. About musical pioneers, he said, "Fletcher [Henderson] was really part of an angelic thing. I wouldn't say he was a man. I wouldn't say Coleman Hawkins was a man, because they did things men hadn't done, and hadn't done before. And they didn't learn it from any man. They just did it. So therefore it came from somewhere else. A lot of things that some men do . . . come from somewhere else, or they're inspired by something that's not of this planet" (56). At moments such as these, Sun Ra is speaking about atypically creative people in metaphorical, hyperbolic terms. Similarly, he told an interviewer that his music was intended for the Creator: "It was really not to be part of the world. . . . I felt that every innovator on this planet was never accepted, whether they were classical composers or otherwise" (122).

In most cases, though, Sun Ra addresses the question of humanity in explicitly racialized and ontological terms. "There now exists a separate kind of human being, the American black man," he said in 1969. "And I should say that he doesn't belong on this Earth" (140). He was reported to have declared during one concert, "This world is not my home. My home is out there" (257). Sometimes, he identified himself explicitly as an angel, "angels conceived of as cosmic musicians" (199). In his poem "When Angels Speak," he wrote,

> When Angels speak / They speak of cosmic waves of sound. (199)

He understood the word *angel* in accordance with its etymology from Greek: *angelos* (messenger), "thus making him a jazz messenger" (314).

Music was Sun Ra's real vocation. Understanding his music is a hazardous exercise because he deals with it on at least three registers: (1) he composes, performers, and records it; (2) he speaks about and theorizes it (not always consistently) in interviews; (3) he makes it a topic of his poetry and song lyrics. Most authors treat these three registers as equivalent, but is there any theoretical justification for doing so? (Obviously, it's easier to say you are an angel than to perform like [or as] one.) Should we assume that music remains the same thing for Sun Ra whether he's speaking about it in an interview or performing it on stage?

I return to those questions later, after a more comprehensive discussion of music and myth within Black critical thought. Before leaving Sun Ra, however, I must note that he occasionally mused explicitly on the topic of interspecies communication from his singular perspective. One of his longtime collaborators, the multi-instrumentalist James Jacson, comments that "Sun Ra wanted to communicate an emotion that people don't know." "Music is the language of the universe. Imagine you meet a creature from another planet. You won't be able to communicate in language, but you will be able to tell something in music. That's what Sun Ra meant when he said his music came from 'outer space'" (120).

In one interview, Sun Ra seemed to suggest that communication with aliens would nonetheless involve reconsidering earthly musical protocols. "Superior beings definitely speak in other harmonic ways than the earth way because they're talking something different," he stated somewhat elliptically (128). But Sun Ra did not dwell on the topic at length, choosing rather to speak in more general terms about the "Black man in the cosmos," as he titled his one-off lecture course at the University of California, Berkeley, in 1971. (It's fascinating to think, from today's perspective, about Sun Ra as a university lecturer. The full reading list from the course has been published online and is worth perusing.)

Two other episodes deserve brief recounting. In the late 1970s, shortly after Sagan's *Voyager Golden Record* had been launched into space, a journalist asked Sun Ra how he felt about not making the final cut. Sun Ra was famous by then, but not on the same level as the luminaries whom Sagan and his team selected: Bach, Beethoven, Stravinsky, Louis Armstrong, and Chuck Berry, to name a few. Sun Ra's response to the interviewer is priceless: "The outer space beings are my brothers. They *sent* me here. They *already* know my music" (351; emphasis in original).

Another fascinating episode happened about ten years earlier, and I include it here primarily because of its relevance to chapter 3's preoccupation with cosmic loneliness. In anticipation of the first-ever moon landing (Neil Armstrong and Buzz Aldrin on Apollo 11), *Esquire* magazine invited famous writers to suggest first words upon landing. Most of the submissions meditated on the loneliness of the accomplishment (275). Others turned the whole thing into an opportunity for expressing Cold War jingoism, touting the US victory over the Soviets. Sun Ra's contribution was something else entirely:

> Reality has touched against myth
> Humanity can move to achieve the impossible
> Because when you've achieved an impossible the others
> Come together to be with their brother, the first impossible
> Borrowed from the rim of the myth
> Happy Space Age to You . . . (275)

Herman Poole Blount grew up a short train ride away from Huntsville, Alabama, where ex-Nazi Wernher von Braun developed the first US space satellite. "*Ebony* magazine reported in 1965 that ten minutes away from the Huntsville research site there were black people living in shacks without water. Wernher von Braun replied in a letter to *Ebony* that the blacks in that town should take pride in being so close to the space age" (xxix). Sun Ra was exposed to the stunning arrogance of men like von Braun his entire life. And it often wounded him—how could it not? But he always believed in the creative and technological capacities of conscious beings, and he understood the paradigmatic significance of the Apollo 11 mission. At least in response to *Esquire*'s invitation, Sun Ra took a "joyful stance as a *witness of alternatives*."[23] Happy Space Age to you, Mr. Ra.

Kapwani Kiwanga

Kapwani Kiwanga was born in Hamilton, Canada, in 1978 and currently lives and works in Paris. Her father is from Tanzania, which is where the name Kiwanga comes from, although in a recent interview, she said that she feels more shaped by her multicultural upbringing in Canada than her familial connections to East Africa.[24] Kiwanga is primarily a visual and performance artist and is at least one generation younger than Sun Ra; her works differ from his in many ways, although she is influenced by them. She came of age at a different historical moment: not the Cold War and the Civil Rights era, but the era of the Washington consensus and the entrance of private funding into space research. Kiwanga is also working at a time when *Afrofuturism* is a household word—no longer restricted to experimental music, the term has been used (if not exactly co-opted) by pop stars and fashion designers. Kiwanga does not see herself strictly as an Afrofuturist artist, but many viewers interpret her work at least partially within that frame.[25] Indeed, some of her work directly references Afrofuturist figures and themes, such as her *Sun Ra Repatriation Project*, an installation from 2009 whose description reads, "The Sun Ra Repatriation Project is an artistic endeavour to repatriate Sun Ra to his homeland: Saturn."

In part because it is a piece of performance art that provides an interesting contrast to Sun Ra, I focus here on Kiwanga's *The Deep Space Scrolls*. In the piece, the artist presents herself almost as a scholar or academic, presenting a lecture accompanied by something like a PowerPoint presentation. It is part of a larger project called the *Afrogalactica Series*, which consists of "three performed-conferences consisting of a live reading accompanied by projected images, video, and sound extracts."[26] Like Sun Ra, Kiwanga engages with African history by uncovering hidden or repressed layers. But, in part because of her personal biography, and in part as a response to Black creative work that appeared during and after Sun Ra's time, Kiwanga focuses less on ancient Egypt. Like Sun Ra, but in a different way, Kiwanga mixes scientific research with fringe views and myth as a way to combat the myopic space research covered extensively in this book. A deep dive into her work will help elucidate Kiwanga's preoccupations.

OCTOBER 30, 2015 (JOHANNESBURG, SOUTH AFRICA; FIELD NOTES)

It is the third day of the African Futures Festival when Kapwani Kiwanga takes the floor. Her presence is strong, controlled; her style is futuristic, suave, elegant. Her hair is sculpted into a rising pompadour, and she wears a monochrome high-neck shirt. Completely deadpan and without a hint of irony, she speaks: "Good evening. My name is Kapwani Kiwanga. I'm a galactic anthropologist from the year 2278. I specialize in ancestral Earth civilizations. I am pleased to be with you today to share some of my findings on my current research on forgotten Earth-Star complexes in terrestrial memory."[27]

Kiwanga's opening gambit is as perplexing as it is alluring: the genre of her talk will doubtless be difficult to define, to anticipate. Not quite academic paper, but not quite performance art either. The African Futures Festival for its part is not at an anthropology conference, a performance venue, or an art gallery—it is a series of multimodal events distributed

> between three African cities and focused on the question of futurity.[28] The context makes interpreting Kiwanga's piece all the more hazardous. Is this high art? (With hindsight, we can compare Sun Ra to avant-garde Fluxus performers, [but] . . . where does *this* piece fit in?) Is it meant to be funny? And who is this person on stage? The program says the artist's name is Kapwani Kiwanga, so is this the same person on stage—that is, the galactic anthropologist? Sun Ra did say that he arrived on Earth from Saturn, after all . . .
>
> "In the past," Kiwanga continues, "intercultural exchange was common between planet Earth and various star systems. These intercultural organisations practiced interstellar travel, as well as cultural, scientific, and technological exchange. This is what we refer to as an Earth-Star complex." In the year 2278, apparently, such complexes are no longer active, but their remnants are "easily accessible in deep space." On earth in 2015, Kiwanga goes on to tell her Johannesburg audience, the residue of formerly robust Earth-Star complexes is in fact evident if one is willing to look beyond mainstream historical narratives. By revisiting the histories of Great Zimbabwe and the Dogon of West Africa, *The Deep Space Scrolls* turns to mythology, ritual, and even conspiracy theory in order to amplify the almost (but not quite) forgotten systems of cultural exchange between humans and extraterrestrials.[29]

One hears echoes of Sun Ra, who devised (and divined) numerous strategies for uncovering hidden connections. Sun Ra's main historical referents were ancient Egypt and the Bible, and if Kiwanga seems to veer in a different direction, note that the title of her piece, *The Deep Space Scrolls*, is, no doubt, a play on the Dead Sea Scrolls, the set of biblical manuscripts discovered in the mid-twentieth century that, because of their differences from other historical manuscripts upon which contemporary Bibles are based, urged scholars to reevaluate key tenets of biblical interpretation. The implication, it seems to me, is that *The Deep Space Scrolls* also compels a devastating reevaluation—this time, perhaps of human history *in toto*.

Like stars themselves, Earth-Star complexes have histories. Kiwanga the galactic anthropologist explains that at some point during a star's life cycle, it begins to "seek out companion societies." Once a stellar civilization has located an appropriate terrestrial partner, "the civilisations begin long term cultural exchange." This exchange is made possible through "star gates," or portals that "enable bilateral Earth-Star traffic." Kiwanga's galactic anthropological research focuses on what happens to Earth-Star complexes at the moment of collapse or destruction—particularly when a star advances to the supernova stage and explodes.[30] In the event of an impending supernova, a stellar civilization is forced to choose between several possibilities. It may

1. migrate to Earth,
2. establish a new community elsewhere in space, or
3. accept the end of its life cycle and perish with the astral calamity.

At the same junction, the Earth society must decide if it

1. accepts or refuses the absorption of stellar fugitives into its society or
2. opts for memory erasure regarding its period of Earth-Star exchange and summarily abandons its social and architectural constructions.

Kiwanga focuses on civilizations in the latter category: namely, those opting for memory erasure. But she is "particularly interested in those cultures which, although having opted for memory erasure, display the persistence of a coded memory." Her lecture performance meticulously documents cases where memory erasure leaves a coded trace.

Consider the Dogon, a West African people who live primarily in Mali and currently number around six hundred thousand. Dogon cosmology is a staple of the anthropological canon, thanks in large part to the work of French scholars Marcel Griaule and Germaine Dieterlen. Griaule's *Dieu d'Eau: Entretien avec Ogotemmêli* (1948) and the co-authored book *Le Renard Pâle* (1965) set the tone for subsequent research, to the point where Mary Douglas once quipped that the Dogon "now seem so unmistakably French, so urbane, so articulate, with such philosophical insight."[31] Indeed, the Dogon appear in Griaule's work as inhabiting an exceptionally, perhaps even suffocatingly, coherent universe. As Douglas comments, "The Dogon [of Griaule and Dieterlen] have created a symbolic structure uniformly embracing their entire universe. The grain of millet in its husk—the human foetus in its placenta—, the world in its atmospheric envelope, are each analogs of the others. The constituent materials and morphology of speech are seen to correspond to those of cereals, of man, of woven cloth, of the whole cosmos. The same intricate harmony of images is drawn down and across from one level of experience to the next."[32] "Reading it," Douglas continues—and here the "it" hovers ambiguously between Dogon cosmology and anthropological texts—"is like gazing through a microscope at a flourishing form of life, confusingly alien and familiar."[33]

The Dogon universe has its origins in the supreme god Amma and is split into two halves: Nommo and Yourougo, also known as the Pale Fox. The Nommo are the serpentlike twin children of Amma; they represent justice, reason, and ritual order—they are redolent of what Jacques Lacan calls "the Law."[34] The Pale Fox is a mischievous figure, a sibling banished from the heavens due to his disobedience and rebelliousness. Speech typically falls on the side of the Nommo; as such, it has a kind of rigidity, what Genevieve Calame-Griaule once termed a "sterile formalism" (*un formalisme stérile*).[35]

The interesting thing is that for the Dogon, at least as described by Griaule and Dieterlen, truth lies on the side of the Pale Fox, and not the Nommo. If the Nommo are purely formalistic, then the Pale Fox is oracular, privy to realities occluded by formal appearances: it "makes mute signs with [its] paws on the sand so that the truth will be revealed."[36] "It is clear," says Douglas, "that the Dogon are not misled by their solidly material theory of speech. . . . Half of their coding system deals formally with knowledge of the world of appearances, and half of it attempts to find short-cuts by another method."[37]

Such is the basis of Dogon cosmology, at least as documented by Griaule and Dieterlen, and as synthesized with a hint of irony by Douglas. But Kiwanga picks up the trail at a more specific juncture, at a mysterious impasse or riddle broached but never elaborated by the French anthropologists. In a short text from 1950, Griaule and Dieterlen observe that the elaborate Dogon cosmology incorporates knowledge of Sirius B, a white dwarf star invisible to the naked eye. They then write, "The question has not been settled, nor even asked, how people who have no technological instruments know the movements and characteristics of practically invisible stars."[38]

In the hands of American author Robert Temple, this single sentence gave rise to one of the most widespread conspiracy theories of the twentieth century. Answering the question that Griaule and Dieterlen barely dared to utter, Temple surmised that the Dogon must have received knowledge of Sirius B from extraterrestrials.[39] Temple's *The Sirius Mystery* was published in 1976, at the zenith of the so-called "ancient astronaut" craze (also known as "cosmonautology" or "paleocontact theory"). Popularized by authors like Erich von Däniken (whose 1968 *Erinnerungen an die Zukunft* was widely influential) and Peter Kolosimo, by the mid-1970s the idea that aliens had built the pyramids of Egypt, the temples of Machu Pichu, and the structures at Stonehenge had gained worldwide traction.[40] Robert Temple proposed a similar theory, this time deploying anthropology to make a case for the influence of alien knowledge on African cultures. In his version of things, the Nommo are not mythical creatures; rather, "Nommo" is revealed to be the name that Dogon people gave to the extraterrestrial beings they encountered.

For Temple and a number of other writers, it was inconceivable that Africans themselves might have developed sophisticated modes of knowledge production. Indeed, between the mid-nineteenth and mid-twentieth centuries, a host of European authors strove to show that Africans were incapable of genius and civilization. Unable to fathom that an African society might have constructed what we today know as Great Zimbabwe, German geographer Karl Mauch (1837–1875) and British archaeologist James Theodore Bent (1852–1897) developed fanciful historical narratives. Mauch attributed the archaeological ruins of Great Zimbabwe to a mysterious lost White civilization (sometimes suggesting that it may be the site of the biblical Ophir), while Bent insisted that a "northern race coming from Arabia" must have been responsible for the impressive stone structures. In a later generation, French anthropologist and priest Henri Breuil (1877–1961)—who worked closely with the South African apartheid government—attributed the paintings he encountered in southern Africa to a "red-haired people (with Semitic profiles)."[41]

Seen in this light, it might seem surprising that some Black artists choose to affirm, rather than refute, Temple's ancient astronaut theory. In *The Deep Space Scrolls*, Kiwanga (in the guise of a galactic anthropologist) restates Temple's hypothesis, stating,

> According to Dogon tradition, the star Sirius has a companion star. The interesting fact is that this companion star remains invisible to the human eye in your present day and was first observed on January 31, 1862, with a telescope by American astronomer Alvan Graham Clark. The Dogon knowledge of this imperceptible companion star, which your scientists call Sirius B, long preceded the first telescopic observation. For many it was inconceivable that the Dogon, having no advanced astronomical tools of observation, could have knowledge of Sirius' companion star. The Dogon Astronomical knowledge became widely known as the "Sirius Mystery."

There are at least four ways to respond to this mystery. One may challenge Griaule's interpretation, as has been done, for example, by Walter van Beek, who does not believe that the Dogon are referring to Sirius B.[42] One could alternatively argue that the Dogon learned about Sirius B from Griaule or another anthropologist. A third option would be to argue that the Dogon did have some kind of technical apparatus or theoretical model that allowed them to discover the companion star. (This option has been dismissed outright by most anthropologists.) Kiwanga takes a fourth option, akin to Temple's ancient astronaut theory, pointing to "Dogon oral traditions" that refer to "a race of people from Sirius called the Nommos [who] visited Earth thousands of years ago." She adds, "The Nommos were amphibious beings that resembled mermen and mermaids."[43] In the conspiracy-laden blogosphere, some propose that Nommos are the model of mermaids everywhere in the world.

A particularly striking aspect of Kiwanga's work is the way she delivers content such as this in a matter-of-fact tone. The effect is uncanny. Kiwanga hardly moves at all during her performances, embodying a scholarly persona. Her contemporary, conference-like aesthetic is belied only by her impeccable futuristic style—from her hairstyle to her slightly shimmering monochrome shirt. Writing about a different but equally speculative piece by Kiwanga, art historian Fanny Curtat asks whether Kiwanga believes what she says. Could it be, Curtat asks, that Kiwanga truly believes in the mysterious ideas that populate her work?[44] After all, Kiwanga certainly acts *as though* she believes in them. With this question in mind, in 2017 I organized an event structured around Kiwanga's work at the Akademie Schloss Solitude in Stuttgart, Germany, where I was a fellow at the time. Taking inspiration from ethnographic biographers, I invited Kiwanga to present her work, followed by a dialogue.[45] I also

FIG. 4.1 Lecture performance by Kapwani Kiwanga in the context of the event "The Social Life of the Universe," at the Akademie Schloss Solitude, May 2017. Photo: Luzi Gross. Reprinted with permission.

invited the sound artist Janine Jembere to produce a creative response to Kiwanga's performance. Composer Bill Dietz and visual artists Dima Hourani and Antoni Rayzhekov also contributed. The outcome was a day-long event—a so-called "performative dialogue"—that we called "The Social Life of the Universe."

As I was planning the event, the ethnomusicologist Peter McMurray suggested to me that, rather than interviewing Kapwani Kiwanga the artist, I should interview Kapwani Kiwanga the galactic anthropologist. When I proposed the idea to Kapwani, however, she immediately responded that she could not to do the interview in the persona of a galactic anthropologist. The problem, she said, is that there would be a confusion between where the performance begins and ends. From this exchange, I learned something important: unlike Sun Ra, who insisted on his actual biographical origins on Saturn, Kiwanga more prosaically conceives of her galactic anthropologist persona as part of a specific performance piece.

Directly following her performance at the Akademie, we did a more traditional public interview, in which I asked Kiwanga to elaborate on her position vis-à-vis her terrestrial and galactic identities. She answered that her work aims to explore aesthetic and political possibilities other than making demands on human equality. Referring to the Memphis sanitation workers' strike of 1968, in which Black men held up signs saying, "I am a man," Kiwanga noted that her artistic practice seeks an alternative path: "Instead of trying to use the same logic, and colonizing oneself by saying, 'Well, we have to prove that we're human,' there's another take which could be like, 'Why don't we think about something else?' Could we be extraterrestrials, could we assume another identity, which breaks beyond and complicates things a bit more."

If, in her guise as a galactic anthropologist, Kiwanga speaks in a certain authoritative manner, her off-stage personality is meditative, undogmatic. She wants to explore different possible identities, she says, complicating the way we think of ourselves. As any researcher knows, posing a good question is the hardest and perhaps most crucial task.

In *The Deep Space Scrolls*, Kiwanga asks hard questions by performatively affirming a version of so-called ancient astronaut theory: the notion that certain civilizations have long been in contact with extraterrestrials. The piece is presented as if by a galactic anthropologist *from the future* finally revealing something *about the past* to us *in the present*. The idea here is that ancient channels of human-alien communication have been obscured and lost, that we in the early twenty-first century have forgotten or repressed the fact that some humans have long interacted with extraterrestrials. This is her way, it seems to me, of rejecting humanism and "thinking about something else."

Addressing her audience during the Johannesburg performance of *The Deep Space Scrolls* in 2015, Kiwanga emphasized that Dogon-Sirius cultural exchange is nearly forgotten in the twenty-first century. And then, to explain "nearly," she added that the trace, the residue of those intergalactic communications can be found in two places. The first is in the stories of conspiracy theorists—not only Robert Temple, but also his precursors (like von Däniken and Kolosimo) and many followers in the blogosphere.

In *The Deep Space Scrolls,* Kiwanga says that there is a second place we can hear the trace of intergalactic communication: Dogon music. Kiwanga does not tell us how this trace can be heard, where in the music's texture and fiber one may hear the echo of something so resolutely other than human. She says only that *she* can hear it, that in returning from the future, and hence in returning our future to ourselves, she can hear a form of civilization that we in the twenty-first century can scarcely imagine. In listening to and for the Nommo, Kiwanga turns in the direction of the Pale Fox, who reveals the truth by creating "mute signs" in the sand.[46] For Kiwanga, those mute signs in the sand speak volumes: they are the ciphers hidden within music.

Alien Trace

Kapwani Kiwanga is not a musician, and her work is not primarily about music. It is somewhat surprising, then, that she turns to music at a critical juncture of *The Deep Space Scrolls.* She attributes to music, as well as conspiracy theory, the capacity to record a trace of human-alien exchange. Again, she does not elaborate in any detail how music does this, and in some sense Dogon music (and music in general) is left hanging as a kind of elliptical provocation. Viewed in isolation, the brief mention of music in *The Deep Space Scrolls* would not merit further discussion, but taking a broader view suggests that much more indeed can be said.

Dogon music has long served as a source of spiritual and psychic support for African diasporic artists. This engagement has been heavily mediated by French anthropology, especially translations of works by Griaule and Dieterlen. Scattered published references and anecdotal evidence suggests that Griaule's *Conversations with Ogotemmêli* was highly influential among members of the Black Artists' Group (a Black artists collective active in St. Louis, Missouri, in the late 1960s and early 1970s). The importance of Dogon cosmology to Black creative practice is evident, for example, in composer and saxophonist Julius Hemphill's 1976 album, *Dogon A.D.* And it is evident, too, in Nathaniel Mackey's extended poem cycle, *Song of the Andoumboulou,* which spans several decades.[47] In what follows, I elaborate on the way that Black theorists, poets, and historians mobilize Dogon music.

One of the most compelling, if under-recognized, writers on the topic is Simone White. Her book, *Dear Angel of Death,* critiques musical and anthropological work about the Dogon in a way that is as impressive as it is demanding. The essayistic, nonfiction portion of her book begins in a somewhat unexpected place: right in the thick of it all, with the first page of the last book by Amiri Baraka (founder of the Black Arts Movement and a primary theorist of Black experimentalism). "I cannot get off the first page of Amiri Baraka's *Digging,*" writes White, who asks us to think of the page as an *image.* She reproduces the page in question in her own book, which I reproduce here as well on the next page.[48] Please do read it before continuing (see fig. 4.2).

White invites us to ponder how "citation makes a visual/verbal bridge that implies a wish to throw down with 'the previous.' . . . The ocular and intellectual stress induced by attempting to read more than one text at a time . . . intensifies textual interplay."[49] What interests me most in the intellectual lineage that *Digging* brings to the surface is the way music is called upon at strategic moments. We are dealing, however, not with musical practice (what musicologists today would call "musicking"—music as a verb), but rather with "the Music": music as metaphor. As White observes, "the great . . . new song" on the first page of *Digging* is "*imagined as heard by Du Bois,* then *imagined by Baraka as an explication of the heard,* the song never actually having been in the presence of either of these two."[50] What, then, is "the Music" a metaphor for? Everything, it would seem:

> Digging Baraka is digging the Black Art of Black Music is belief that black soul is movement outside the word of white supremacy's denial of black existence, which contains the possibility of spontaneous pleasure, which is beauty. Digging is a theory of the time-space boundary of black thought that hypothesizes a way to get to the outside of white totality, of white supremacy's word, which always, always, refuses the capacity of black people to move in a motivated way together through time-space, toward the end of white supremacy, which is freedom, or also, the end of time.[51]

The entire time of historicity is literally packed into this passage. White supremacy is the denial of Black existence, says Simone White, a refusal of the capacity of Black people to move in a motivated way—this

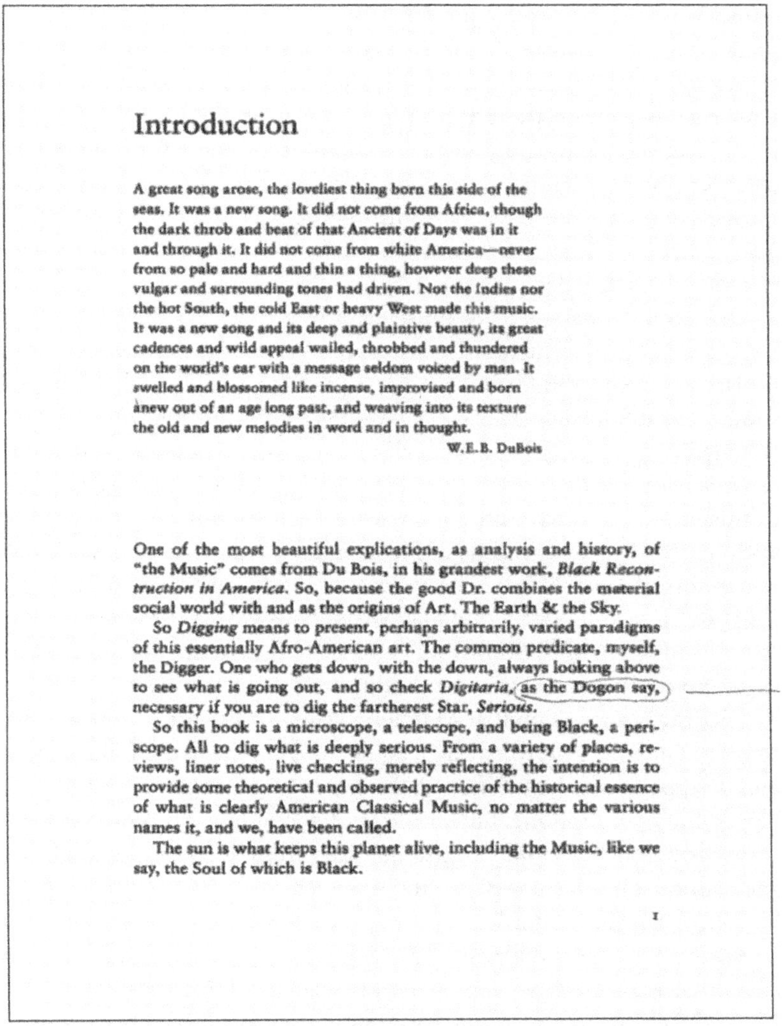

Introduction

A great song arose, the loveliest thing born this side of the seas. It was a new song. It did not come from Africa, though the dark throb and beat of that Ancient of Days was in it and through it. It did not come from white America—never from so pale and hard and thin a thing, however deep these vulgar and surrounding tones had driven. Not the Indies nor the hot South, the cold East or heavy West made this music. It was a new song and its deep and plaintive beauty, its great cadences and wild appeal wailed, throbbed and thundered on the world's ear with a message seldom voiced by man. It swelled and blossomed like incense, improvised and born anew out of an age long past, and weaving into its texture the old and new melodies in word and in thought.

W.E.B. DuBois

One of the most beautiful explications, as analysis and history, of "the Music" comes from Du Bois, in his grandest work, *Black Reconstruction in America*. So, because the good Dr. combines the material social world with and as the origins of Art. The Earth & the Sky.

So *Digging* means to present, perhaps arbitrarily, varied paradigms of this essentially Afro-American art. The common predicate, myself, the Digger. One who gets down, with the down, always looking above to see what is going out, and so check *Digitaria*, as the Dogon say, necessary if you are to dig the fartherest Star, *Serious*.

So this book is a microscope, a telescope, and being Black, a periscope. All to dig what is deeply serious. From a variety of places, reviews, liner notes, live checking, merely reflecting, the intention is to provide some theoretical and observed practice of the historical essence of what is clearly American Classical Music, no matter the various names it, and we, have been called.

The sun is what keeps this planet alive, including the Music, like we say, the Soul of which is Black.

1

FIG. 4.2 The first page of Amiri Baraka's *Digging*. Reproduced from Simone White, *Dear Angel of Death* (Brooklyn: Ugly Duckling, 2018), 77; Amiri Baraka, *Digging: The Afro-American Soul of American Classical Music* (Berkeley: University of California Press, 2009), 1. Reprinted with permission from the University of California Press and from Ugly Duckling Press.

refusal is what scholars like Jared Sexton and Frank B. Wilderson III call social death.[52] According to the lineage signaled by Baraka, there is, however, a way to get outside that totality, and that way is musical excavation. How else would Sun Ra return to his place of origin, Saturn? How else would Kapwani Kiwanga, the galactical anthropologist from the future, discover traces of human-alien cultural exchange? How else, other than through digging? The galactical anthropologist is more of an archaeologist, someone who digs not into the earth—not, even, into Earth—but into the Music as a decipherable archive of it all.

One explicit statement of music as a way to access the outside can be found in Nicole Mitchell's short essay, "What Was Feared Lost." As a flutist, composer, and former chairperson of the Association for the Advancement of Creative Musicians (AACM), an important organization, Mitchell's relationship to music is clearly more than metaphorical. In this sense, she differs from many of the figures discussed in this section (and in this book), who have little if any experience performing music. That said, her written descriptions of music in that essay take on a decidedly metaphorical tone. She begins by describing the world as a "cramped shoebox," a cage-like structure that we previously believed was meant for "slaves and women," "but now we've come to realize that we are all stuck in it—no exception."[53] However—and here one can almost viscerally anticipate the intervention that Mitchell makes—music functions as a way to get outside the box, to redesign an opening, or make the walls tremble and disappear. Mitchell describes the experience of traveling aboard the "spaceship of saxophone, flute, cello, and percussion."[54] "With Arveeayl Ra's meditative bells and truthful questing from Tomeika Reed's cello," she continues on the next page, "there was a beyond. She saw through the keyhole—another realm!"

Mitchell's essay is relevant to the concerns of this chapter for at least two other reasons. First, she invokes the Dogon. What lies outside the walls of the shoebox is "untouchable and unseeable," she writes, "but it's real." She continues, "You know, better than that, if you practice your connection to the stars enough, you can go anytime you please. *The Dogon mastered it*, but the shoebox architects sold kids on rap videos over skygazing."[55] In this passage Mitchell valorizes Dogon knowledge, contrasting it with rap videos. As Simone White observes (and as we will see in more detail in a moment), Mitchell's dismissal of rap is symptomatic of a particular strain of Black critical thought.

The second relevant aspect is implied in the essay's title. For Mitchell, music preserves the residue of what lies outside the box, that is to say, of what was feared lost—what was feared but is actually not lost, can never be lost, because music will hold on to that trace. She writes, "My secret is this. The shoebox world seems like a vast maze, but in fact, it is just a tiny cell on the blue marble floating in the endless dark maroon cloud. The dark unnamable is much, much bigger, and therefore, much harder to know, but it's all love. *It's what was feared lost*."[56] In a style typical of her writing, what follows is a seemingly random example of how the secret of the unnamable is revealed: "Our world, it's so full of green. It's a playground for catching gems of truth, like Zorn's Stone."[57] Here, Mitchel positions John Zorn's legendary East Village music venue as a microcosm of the world's verdant, unnamable reality. Music as residue, music as trace: this is a key theme. For Mitchell, a trace of the lush world outside our shoebox existence can be heard in music. And Kiwanga the galactic anthropologist explains that the trace of Dogon-Sirius cultural exchange is preserved—somehow—in Dogon music. There is a residue of the otherworldly in music, in other words, one that Sun Ra also held onto.[58]

One consequence of the music-as-trace concept is the inseparability of social from cosmic concerns. In a way, music is registered not only as a trace but as a kind of audible microcosm of hidden cosmic realms. As Baraka says on the fabled first page of *Digging*, the "good Dr. [Du Bois] combines the material social world with and as the origins of Art. The Earth & Sky." According to Sun Ra's longtime musical associate, Walter Miller, Sun Ra also often paired "outer space with earthly matters."[59] Continuing this intellectual lineage, Nathaniel Mackey concurs: "Social estrangement is gnostic estrangement and the step from Satchmo's 'height of the trumpet' to Sun Ra's 'intergalactic music' is neither a long nor an illogical one."[60] Music, in this rendering, is "the poetic dwelling of blacks," music names "how black people fill up space between earth and sky (materially, socially)."[61]

Within the tradition of Black critical thought I am addressing here, Music forms precisely the link between social materiality, Earth, and Sky. That mobilization, in turn, propels itself through what Simone White terms a "postmodern recuperation of Dogon cosmology."[62] As Baraka says on that first page, to find that hidden star (Sirius B), you need to "get down . . . with the down," you need to move with music and start digging.

According to the writer Frank B. Wilderson III, all other social groups have recourse to a space and time of plentitude—Israel for the Jews, the Joseon Dynasty for South Koreans, and so on. The condition of Blackness is by contrast one of natal alienation, which means the total absence of any possibility of emancipatory return.[63] In Wilderson's theorization, which is obviously controversial, Blackness is understood as a "paradigmatic" social position rather than a cultural form, and it's because of this sense that he flat-out rejects the possibility of emancipation through cultural reclamation, such as in the case of Alex Haley and *Roots*. Wilderson writes, "Temporal capacity (the heritage of historiography) permits one to think 'time can be *redeemed*.' For time to be redeemable, time must have been, at some historical moment, deemed. But the most coherent temporality ever deemed as Black time is the 'moment' of no time at all on the map of no place at all: the ship hold of the Middle Passage."[64] What kind of a dwelling can music provide? Wilderson would say that it can provide none at all, and in this sense he is at odds with most of the thinkers discussed in this chapter. As a self-avowed Afro-pessimist, Wilderson would say that Black people will only have a genuine space of dwelling when a new world comes into being—music might be a powerful tool, but it cannot offer a sanctuary in a world as violent and racist as this.[65]

I lived down the street from The Stone between 2010 and 2012 and visited it countless times during that period—The Stone was my favorite music venue of all time. The Zim Ngqawana, Henry Grimes, and Andrew Cyrille concert on November 30, 2010, was perhaps the best concert I've ever attended. But it is one thing to love music and to love a venue, and another to freight a music venue with cosmic import. By so investing it, Mitchell's writing, if not her performance, resembles Baraka's.[66] It would seem, in fact, that the metaphor of "the Music" as used by Mitchell, Baraka, and others at times comes close to the metaphysicalization of music augured by nineteenth-century European aesthetics.

From a certain perspective, the musical metaphysics of Mitchell/Baraka and someone like John Lilly form two sides of the same coin. But the fact that Lilly and Mitchell draw on similar (and similarly problematic, at least in my view) intellectual resources should not shroud the different uses to which they put concepts like "the Music." Lilly, at least in the decades surveyed by this book, was an opportunist engaged in cruel practices; Mitchell, for her part, is a teacher and mentor to young performers and an active participant in the Association for the Advancement of Creative Musicians, an organization in the service of creatively and economically empowering Black musicians. Does noting these differences mean that I am being moralistic or that I am "judging" Lilly? No, it does not.[67]

At least at this point in the chapter, however, my focus rests on the conceptual framework used by Black musicians, poets, and activists. On this score, Simone White takes issue with the way Music is rendered metaphysically by Baraka and his followers. She worries that the notion is too metaphorical, and that it elides the way music is practiced and experienced by most Black people today, especially young Black people.

(She takes issue with the insistence on experimental or avant-garde music, when most young Black people are more interested in popular music. Recall Nicole Mitchell's derisive remark about shoebox architects and rap videos.) Turning away from Dogon-inspired, experimental musical mysticism, she invites us into the world of trap music, a genre of Southern rap that emerged in the early 2000s. The lyrics of trap are almost the opposite of Afrofuturism, turning the gaze from the heavens and galaxies beyond to the club, to violence, and to narcotics. Musically, trap makes use of repetition (often excessive repetition), and is often identifiable by rapid, electronically-generated hi-hat hits—at times so fast that the hits begin to blur together, almost forming a continuous frequency. The production value of trap music is usually high, its producers virtuosi of studio technique, effortlessly controlling eccentric cymbal patterns, swooping bass drums, and oscillating metrical divisions.

In opposition to the mythical, cosmic, and metaphysical (non-)space of a figure like Sun Ra, White considers the trap musician's "ability to develop a deranged and *hyper-cognitive* understanding of her location on a grid of recordings—textual, somatic, spiritual or what have you. The theoretical place of music here is, at best, indeterminate, conditional upon a person's individual capacity to cognitively and spiritually disengage from certain apparently inescapable 'material conditions.'"[68] Jettisoning the metaphorical and cosmic space of "the Music," White turns her ears toward the way musicians engage the world as it is around them, participating in something closer to immanent critique than cosmic travel. She says that contemporary trap music is "true in its bewildering rejection of its own status as the start of a certain kind of radical thinking about getting out or outside."[69] This is not a music outside the shoebox world, but one inside it, intimately inside and pressing at the seams.

Her analysis is elliptical, however, and short. What she wishes to show, above all, is the way that artists like Vince Staples and Future reject the White desire to have a world (indeed, to possess the entire world), choosing instead to participate in what she terms a covenant: "The covenant that makes our 'thing' a thing that can be apprehended is the promise to stay together in the absurd or 'foolish' space of the self that accepts itself as black, surrounds itself with blacks, a promise that is made and renewed by a self in 'constant improvisational contact' with other facets of the world, organic and inorganic alike: earth and sky and the human and his air."[70] Only by hewing to that place do musicians make contact with what is outside and beyond them—but only indeterminately, and without any guarantees.

Commenting on the 2015 song, "I Serve the Base" by Future (a popular rapper out of Atlanta), White emphasizes the extent to which the sound and lyrics of the song aim for anything but cosmic ambitions. "I serve the base [i.e., drugs], I serve the base," Future repeats. White praises the nihilism and endless repetition of the song, the way it produces a "narcotized kind of peace."[71] The song is a giving up, a relinquishment, of everything, "the call to recede into the persona of whatever it is one serves."

About Vince Staples, a young musician from California, White emphasizes a similar kind of nihilism:

> There's an emotional and ideological grid that Vince Staples has access to, but hasn't built. That grid is structured by the axes of being/living as black and the practice of representing blackness. Vince hasn't built it—and it's not his fault—yet he plays or practices a black game which is also the game of *repeating what we already know, repeating facts to which it is possible to nod our heads, facts such as "the music and the people is the same."* We watch him work a terrifying manipulation of ideological and emotional materials that are fundamentally out of his control and so drop him, Emily [Dickinson] says, BONE by BONE. That is what it is to look, if you look, into the abyss, the organized inside space of blackness.[72]

In a way, Simone White inverts the terms of Afrofuturism, pointing to the engagement and manipulation of "materials" leading to a kind of emptying out. While she does not address humanism directly, an urgency to disrupt the liberal-humanist conceit is evident everywhere in her writing. And even though she does not say so, she seems to suggest that Whiteness is equivalent to what Sexton would call the "desire for a world," while the musicians she celebrates declare nihilistic detachment. Where would that detachment leave them? Where would it take them? "To Saturn," Sun Ra might respond. Simone White offers a different, and in some sense more prosaic answer.

The Humanist Dilemma

In one extreme formulation, Frank B. Wilderson III argues that comparing anti-Black racism with other forms of discrimination "erroneously locates Blacks in the world—a place where they have not been since the dawning of Blackness."[73] This theorization mobilizes a concept, a highly controversial notion of world, but one that approximates Afrofuturism, and from which we can learn.[74] But, considering Sun Ra's talk of outer space, can we assume that the world is the same thing as the Earth? If not, where does the one end and the other begin?

To sharpen our definitions, I turn back to some foundational philosophers of modernity, beginning with Kant. Recall that for Kant the extraterrestrial is a being that lives outside of the Earth but that nonetheless inhabits a similar (or even the same) world as terrestrial humans. This doctrine was known as the "plurality of worlds," and referred to distant planets filled with beings analogous to ourselves with whom we would likely be able to communicate—or, at the very least, compare our rationality.[75] Also in a way typical for his time, Kant saw the animal as a being that lived on Earth with humans but that nonetheless inhabited a fundamentally different world from that of terrestrial humans. In short, the alien lives elsewhere but is in our world, whereas the animal lives alongside us but in a different world.

In post-Kantian European philosophy, the question nonetheless remained whether animals truly live in different worlds or whether they live in something else entirely—not quite a world, but some lesser habitation. More than a century after Kant, Heidegger would develop a schema according to which stones are worldless, animals are poor in the world, and humans live in a world of their own making.[76] But even in this neat schema, we cannot forget the catch: for Heidegger, as for much of the European philosophical tradition, not all humans are equally capable of producing a world.

European modernity has often thought of itself as having a world—of rationality, meaningfulness, history, and so on. But those positive assertions have always relied on an opposite. Hegel's philosophy of European-led historical development was bolstered, for instance, by his claim that Africa is "outside history."[77] Because Africa did not develop in terms recognizable to Hegel (or rather, because he was unable to recognize the terms of Africa's development), he thought that Africa was not part of history, and that it was not part of the human world. To Hegel, Africans *were* aliens of a sort: not in the sense of living on Mars, but in the sense that they did not partake of human existence, which consisted of great world-historical individuals like Socrates or Julius Caesar.

Sun Ra responded in two different, yet overlapping ways. First, he sought to uncover robust African civilizations from under the mirage of White history. Second, he turned to myth as a way of refuting the terms of Hegelian logic: if Hegel was interested in the teleological development of national histories, Sun Ra focused his attention on the more profound mythical state of being that preceded the time of history.

Authors like Wilderson are skeptical of Afrofuturism, however. Wilderson argues that while Afrofuturists think they are imagining a world outside the one we inhabit, they are merely projecting ideas from our world: “You cannot imagine another world inside of the episteme in which you live. That’s where I differ from the Afrofuturists. They’re using linguistics and semiotics to say that we’re out of the realm of absolute abjection, and that’s just religious thinking. It moves from analysis to idealism without showing the move.”[78] Media theorist Kara Keeling criticizes Sun Ra in basically the same way:

> Sun Ra’s solution in *Space Is the Place* to the violence and humiliation of US race relations is to give Black folks a world they can own. Sun Ra seems unconcerned about the specter of African American and Black complicity in a settler colonial project when he advocates for a spatio-temporal rupture in Black consciousness sparked by his musical vibrations and profound enough to transport Black people to another planet. Afrofuturist narratives that advocate for colonizing another planet raise (and less often consider, and/or offer, speculative strategies and solutions to) the ethico-political issues that have attended anti-Black settler colonial societies.[79]

Keeling concludes, “From within the logics of existing possible worlds and the range of possible trajectories into the future that they currently make perceptible, a Black future looks like no future at all.”[80]

Taken at face value, Sun Ra’s will to escape and terraform (or perhaps vibraform) another planet indeed sounds like a settler colonial project. But I think that Keeling is taking Sun Ra too literally—or, better stated, too directly. I contend that, at least to some extent, Sun Ra and Kapwani Kiwanga avoid the pitfalls outlined by both Keeling and Wilderson.

Superficially, at least, the final scene of *Space Is the Place* (where Sun Ra gets on a spaceship leaving planet Earth) would, vis-à-vis Keeling, seem to “advocate for colonizing another planet.” The first scene (where Sun Ra is walking around a strange planet with weird vegetation) would seem, vis-à-vis Wilderson, to point at some fantasy place, without showing how it offers anything different to the violent and racist world that Sun Ra was forced to live in. Indeed, Wilderson is trained as a film critic, and he has argued that, from a narratological perspective, Black subjects can never genuinely appear. “Can film tell the story of a sentient being whose story can be neither recognized nor incorporated into Human civil society?” he asks—rhetorically, of course, since his answer is a resounding no.[81] Wilderson says the same thing about technologies of filmmaking and film viewing: “The Black can be placed *on* film but cannot be positioned *within* the frame; . . . the Black bears witness to a space and time she cannot enter: the space and time of the world.”[82] This is close to the description provided by Ralph Ellison in *Invisible Man,* when the protagonist of the novel opines that all the White people he knows "were very much the same, each attempting to force his picture of reality upon me and neither giving a hoot in hell for how things looked to me. I was simply a material, a natural resource to be used."[83] The message of the novel is that being invisible means appearing to be however others (in this case, White people) see one to be, without ever being permitted a substance or perspective of one's own. (In Simone White's view, musicians like Future and Vince Staples radicalize that invisibility, receding "into the persona of whatever it is one serves.")

But consider a different scene in *Space Is the Place.* At a community center in Oakland, California, Sun Ra materializes out of nowhere. Young Black women and men are shooting pool, playing ping-pong, and singing, when they are interrupted by his sudden presence. Startled, a young woman asks: "Is he for real?" And to this Sun Ra responds, "How do you know I'm real? I'm not real, I'm just like you. You don't exist in this society. If you did, your people wouldn't be seeking equal rights. You're not real. If you were, you'd have some status among the nations of the world. So we are both myths. I do not come to you as a reality, I come to you as the myth. Because that's what Black people are—myths."

In his answer, Sun Ra forces a shift of register: when the young people ask, "Is he for real?" they obviously are not asking a question regarding the nature of his existence. But Sun Ra's response comes swiftly: "I'm not real." A few seconds later, he shifts the register again. Sun Ra talks about being a "presence sent . . . by your ancestors," saying that he has come to take some Black people from Earth back with him. "Are there any whiteys up there?" one young person asks. Sun Ra answers, "They're walking there today. They take frequent trips to the Moon. I notice none of you have been invited." In his response, Sun Ra mixes two notions of world: the world as a space of human social reality, and the world as terrestrial habitation.

Sun Ra claims to exist not as a reality but as a myth—even when he seems to be standing firmly on the ground in Oakland, California. And he claims the same for all Black people. What about his idea to start a society without White people? As he says in the beginning of the film, this would be done in the hope of seeing what Black people "can do on a planet all of their own"—there is no suggestion of magical or religious thinking here, only the vague hope of something radically unknown and unknowable. While I don't think that Sun Ra is culpable of colonial thinking, it would be disingenuous to deny that Wilderson's critique of Afrofuturism does not apply at all: Sun Ra was indeed a "religious thinker," unabashedly so. But Sun Ra was also more and less than that, as many of the examples presented in this chapter show.

What about Kapwani Kiwanga? Is her work susceptible to charges of religious thinking and idealism? Kiwanga's work is not easy to analyze, and is open to many possible interpretations; here I consider three. It would be possible, for example, to view her work in primarily anthropological terms, focusing on Dogon cosmology and in particular on the Nommo. On this reading, Kiwanga may be thought to participate in a decolonizing gesture by affirming ancient African belief systems. Interestingly, the Nommo (beings that represent speech and reason) are typically depicted as mermaids or fishlike creatures, and in this sense we are reminded of Xhosa cosmology, in which similar beings play a role.

Understood on these grounds, *The Deep Space Scrolls* can be viewed as Kiwanga (re-)connecting with African spiritual forces—and, certainly, it would be difficult to *not* interpret the performance to be doing that, at least partially. I described the influence of Dogon music and cosmology on Black musicians and writers in the previous section. Much of that influence stems directly from *Conversations with Ogotemmêli,* the 1965 English translation of Griaule's *Dieu d'Eau: Entretien avec Ogotemmêli* of 1948. *Dieu d'Eau* was important primarily because of its widely available English translation of 1965, whereas *Le Renard Pâle* was first published in 1965 and received an English translation only in 1986. (The fact of their relatively late translation dates also explains why Sun Ra paid scant attention, if any, to the Dogon.) That English translation was produced by "leaders and members of the Black community in Minneapolis," who "organized themselves as an ad hoc committee to seek funding."[84] Incidentally, when the book was finally published by Afrikan World Books, the ad hoc committee gave special thanks to "Dr. Frank B. Wilderson, Jr., Vice-President for Student Affairs and Professor of Educational Psychology at the University of Minnesota in Minneapolis," who was instrumental to the success of the project. This Frank Wilderson is the father of Frank Wilderson III, to whom I've referred several times in this chapter.

Dogon cosmology also makes an unexpected appearance at the end of Achille Mbembe's recent book, *The Critique of Black Reason.* Written by one of today's most prominent postcolonial and political theorists, *The Critique of Black Reason* alludes, of course, to Immanuel Kant's three critiques, but also moves in a radically new and different direction. In the epilogue titled "There Is Only One World," Mbembe begins, "The birth of the racial subject—and therefore of Blackness—is linked to the history of capitalism. Capitalism emerged as a double impulse toward, on the one hand, the unlimited violation of all forms of prohibition and, on the other, the abolition of any distinction between ends and means. The Black slave, in his dark splendor, was the first racial subject: the product of the two impulses, the most visible symbol of the possibility of violence without limits and of vulnerability without a safety net."[85] But if the Black slave was the *first* subject to experience limitless violence, today that violence has become generalized. "The emergence of new imperial practices is then tied to the tendency to universalize the Black condition," says Mbembe, referring to the "becoming Black of the world."[86]

What interests me here is the way Mbembe mobilizes Dogon cosmology against what he sees as the violence of colonial and new imperial logics. In making a case against a world order in which "no region . . . is spared from the logics of the distribution of violence on a planetary scale,"[87] Mbembe turns to the Dogon: "Among the ancient Dogon people, the unending labor of reparation had a name: the dialectic of meat and seed. The work of social institutions was to fight the death of the human, to ward off corruption, that process of decay and rot. . . . It was therefore a way of returning to the idea that, as long as the work of reparation continued, life was an imperishable form, one that could not decay."[88] This is a fascinating moment in the history of political thought. Mbembe positions Black reason as the fourth critique, overcoming the three by Kant (*The Critique of Pure Reason, The Critique of Practical Reason,* and *The Critique of the Power of Judgment*). But where does Mbembe ultimately find Black reason? Where else than in Dogon cosmology? Much more could be said of Mbembe's thought in relation to this book's concerns; for now, I want only to illustrate the continuing recourse to Dogon cosmology by contemporary theorists and artists. And I want to emphasize, as well, how this recourse is inevitably mediated by French anthropology of the mid-twentieth century. Although he does not say so explicitly, it would seem that Mbembe has Dominique Zahan's *La Graine et la Viande: Mythologie Dogon* (1969) in mind when he refers to the "dialectic of meat and seed."[89]

A second way of understanding Kiwanga's work is in the context of speculative fiction and fringe science. This register is somewhat troubling, because it potentially verges on conspiracy theory. While Robert Temple's *The Sirius Mystery* is not pernicious on its own, theories of alien contact often seep into and provide fodder for more insidious ideas—for example, the so-called "reptilian" theory promulgated by David Icke, which holds that the Earth is being controlled by shapeshifting reptilians who have the ability to take on human form. Icke's conspiracy theories often dovetail with classic anti-Semitic tropes, and many have mobilized his ideas to claim that the reptilians are in reality a cabal of global Jewish bankers. (Alice Walker recently came under scrutiny for promoting Icke's ideas.) In our current climate of anti-vaxxer rhetoric, "fake news," and QAnon, conspiracy theory is more dangerous than ever.

Kiwanga's *Deep Space Scrolls* is open to these readings—her work is indeed susceptible to being interpreted as romanticizing Dogon cosmology or as feeding into conspiracy theory. But other interpretive possibilities exist. Kiwanga carefully maintains the border separating performance and everyday life, and the content about Dogon cosmology and the Sirius Mystery in her work should not be taken literally as content that she is affirming or advocating. She is not, like some Afrofuturists, presenting a future fantasy world in which oppressed people are finally free. One possible interpretation, then, is that she is participating in the same kind of epistemological critique as those critical of Afrofuturism.

Kiwanga makes it difficult for an audience member to follow the sequence of events in her performances or to figure out how they should be experiencing the event. It seems funny (hilarious at times), but Kiwanga speaks with cool authority. The subject matter is at times tragic (the history of colonialism, for example), but then Kiwanga will introduce conspiracy theory into the mix. The different registers of inside/outside disorient the spectator: we are invited to hear the words of a galactic anthropologist from the future speaking about the past (as well as, I might add, the mythical past) in a heady mix of fact, fiction, falsehood (the definitely not true), and speculation (the possibly true which is nonetheless unverifiable). Considering that even "experts" are ill-equipped to deal with many questions concerning the existence of aliens and the nature of life, extraterrestriality remains a zone—if not exactly of free imaginative play—then of speculative possibility.

I have presented Kiwanga's work here in granular form, and with the benefit of having access to the text that is the basis of *The Deep Space Scrolls*. But audience members only get to see and hear the piece once, unfolding quickly and in real time. (Kiwanga does not permit recordings of her work; her performances are not available anywhere online, on YouTube or anywhere else.) To experience *The Deep Space Scrolls* is at once thrilling and disconcerting—it is akin, in a strange way, to the definition of avant-garde performance as denying a sense of coherence.[90] Even though the audience sits directly facing Kiwanga, the coherence of the spectacle is indeed abandoned, not in a simple spatial or temporal way, but because its members do not know where they are located in a more fundamental sense. Although there are obvious motifs in Kiwanga's work, when I spoke with audience members after her performances in Johannesburg (2015) and Stuttgart (2017), interpretations of the event varied widely. Of course, critics can and will continue to come up with coherent interpretations of her work, but one important thing about her

Afrogalactica Series, I suggest, is the way it resists criticism. The viewer-listener is "displaced from . . . authority," unable to discern the boundaries between performance, historical exposition, speculative intervention, and myth.[91]

I borrow the notion of the avant-garde as a practice that displaces spectatorial authority from the art historian Mike Sell. Sell formulates that hypothesis particularly in relation to Fluxus (he focuses on Allan Kaprow), but his formulation is further radicalized by Afrofuturist artists like Kiwanga because of the vexed history of Blackness and its appearance within the long history of Euro-American representational systems. As Ellison elucidates in *Invisible Man,* representations of Blackness are often simply projections of whatever White viewers want to see (this is why Sun Ra says, "I do not come to you as a reality. I come to you as a myth."). The history of Africa, likewise—at least as documented by White scholars over the past several centuries—is also in many ways the history of fantasy projections. As Mbembe puts it, "On a phenomenological level, the term [Black] designates not a significant reality but a field—or, better yet, a coating—of nonsense and fantasies that the West (and other parts of the world) have woven, and in which it clothed people of African origin."[92] The "Black Man," he says, "exists where he is not thought."[93] One important aspect of Kiwanga's work is the way she plays with the boundaries between visibility/invisibility, audibility/inaudibility, reality/myth. She does not imagine a world outside of our current one as much as she reflects already-existing mythic fantasies back onto her audience.

On a different register, Kiwanga's work is explicitly about interspecies communication. The main thrust of *The Deep Space Scrolls* is that the Dogon were indeed in contact with an extraterrestrial species on some distant planet, but that the import of that contact will only be fully understood in the future. On this view, the Dogon (and perhaps other African civilizations as well) have long developed the capacity for interspecies and interstellar communication in a way that White scientists have consistently failed to do. The implication here, I think, is that science's protocols have obscured rather than sharpened our capacity for interspecies and intergalactic communication—conceiving our human presence in the universe in colonial terms obfuscates possible communion with "maximal strangers."[94] In other words, only because the Dogon have not historically shared the values of White, European "Man"—hyper-rationalization, colonial expansion and development, and so on—are they receptive to interstellar messages. Only because of their distance from the Human can the Dogon adequately hear the cosmos beyond our planet.

An Alien Listening

Throughout this book, I have remained critical of attempts to grant music too much power: power to cross the human/animal divide, power to show us the beauty of whales so we stop killing them, power to take us into the heavens. I have argued repeatedly that music can and does produce metaphysical imaginaries, but that it is, above all else, a semiotic system. Simone White critiques the work of Baraka and others, arguing that they are less interested in actual music than in "the Music," a metaphorized set of hopes and desires. Her turn to trap music is laudable—after all, that is the music that a lot of young Black people listen to, so it deserves to be taken seriously. But I also think there's another way to approach the experimental Black music that Simone White turns away from. As a step in this direction, consider John Szwed's final appraisal of Sun Ra at the end of his brilliant biography. I quote him at length:

> [Sun Ra] might also be remembered as a composer in the grand tradition, one who was driven by a hunger for totality that only music could express. To him poetry, dance, and music were linked together as arts of the highest order, and music—especially instrumental music—was the most immediate means for engaging the emotions with a higher reality. Music could provide a metaphysical experience through which one could enter the sublime, and come to know the cosmos. He understood music to be a universal language, and something akin to religion. Music could convey more than feelings about phenomena; it could express its essence, and thus could disclose secrets of nature not available to reason, secrets which reveal the true nature of the world. Sun Ra saw that music symbolized the unity in diversity that is the cosmos, and the big band was his space vehicle, African-American aesthetics his culture-synthesizing principle. He was the bandleader as prophetic leader, the music arranger as arranger of the world.
>
> Such a program was enough to qualify him as a European romantic. But his was also an African-American romanticism, its goal a collective metaphysical experience. Though like the Europeans he began with a conception of the body as an individual instrument sympathetically vibrating to music—thus linking the body to the celestial, the harmony of the spheres, the experience of the flight of music, the hovering of music in space—he saw it not as private experience, the artist alone cut off from other artists. In Sun Ra's romanticism, art is capable of constituting a community which mirrors the universe, an artist's vision of the black sacred cosmos.[95]

What I like about this description is that Szwed insists on the grounded, practical aspect of Sun Ra's work. To reiterate an idea from earlier, Sun Ra approached music in multiple ways, and not only as a musician. Sometimes there exists a gap between the descriptions of music in his poetry and in interviews, and another gap between both those kinds of descriptions and the way his music sounds. Sun Ra might pontificate about the ability of music to transmolecuralize people into outer space, and follow that with a composition that sounds like a flurry of saxophone noodling. Other compositions seem to push in the musical direction of something more cosmic, however, such as his "Friendly Galaxy," which deploys a three-note keyboard motif in a way that constantly shifts the accent so that there's a constant tension as to where the first beat of the measure is. The listener feels disoriented, foreground and background constantly reversing and pulling away from each other. On top of that perceptually shifty riff, various musicians improvise in a way that displays their individual creativity as well as their astute listening within the collective. Add to these dimensions the use of new musical technologies with the capacity to create wild, novel timbres, and you begin to get the sense, as a listener, of the otherworldly within the music. What one hears is the endeavor to forge a new kind of sign from the other side (one might say from the future or from "the Other") of music: a panoply of noise, fragments, negation, technological excess, and so on.

In closing this chapter, I want to suggest that there is something productive about the non-alignment (and occasional alignment) between registers: between the music and "the Music," as it were. At the same time, to examine Sun Ra's music by considering only what he says about it is to miss something fundamental, something precious. Although Sun Ra only occasionally mused on interspecies communication, his life's work—and, indeed, he treated his own life as a kind of creative labor—sought to undermine ways of living as a human that make connections beyond ourselves impossible. He was interested, in other words, in finding ways to open humanity up to the world around us. Sun Ra was quoted as saying, "I was born with x-ray ears. I can hear all these things you humans can't."[96] Or, as he stated elsewhere, "Jazz . . . is a communication point to somewhere else."[97]

Rather than think of his music as a performance and then consider only the content of his spoken words, we might instead understand everything he did as part of the same project to listen otherwise. And the same can be said of Kapwani Kiwanga. What will we know about humans' capacity to communicate with extraterrestrials in two centuries? Will we even make it that far? It seems never to have occurred to Great Filter theorists that they themselves might be the filter. What kind of thinking leads a civilization to destruction? What kind other than the mongrel expansionist solipsism that rules the day? Kiwanga's work begins to pry open this domain, creating a confrontation between African history, astronomy (past and future), and conspiracy theory in a way that resists comprehension in the simple sense.

What I am suggesting, ultimately, is that Sun Ra and Kapwani Kiwanga are a lot more like Future and Vince Staples than one might assume. While the first two don't affirm the radical nihilism that Simone White attributes to the second two, they seem to engage with rather than try to escape their "inescapable 'material conditions,'" as she puts it. Despite invoking the alien, both Sun Ra and Kapwani Kiwanga are performers first and foremost, twisting and cracking the world of sound and images from within. They work with a bewildering array of texts, sounds, histories, and possible futures, chipping away at any coherence an audience member might expect. They seem, in fact, to embody Baudrillard's three theorems, which I paraphrase here as follows: to make the world more enigmatic, to take a frenzied view of the world, and to guard against theory overshadowing the world it is supposed to explain.

As stated earlier in this chapter, neither Sun Ra nor Kapwani Kiwanga provide an obvious guide for interspecies communication. Rather, they show just how much needs to be cleared away before a further step in that direction might ever be possible. They don't do this, in my view, by using music to directly access some verdant, starry beyond (to offer a bit of a mixed metaphor). Rather, they participate in a veritable *Durcharbeitung* (working-through) of colonial history, which is also the history of humanity's relationship to space—in the double sense of spatiality and of outer space. To adapt Lyotard's words, "The avant-garde's unremitting work of anamnesis has, for a hundred years, saved the honour of thought, if not of humanity, without compromise and everywhere. It is not enough, but it is something."[98]

To save "the honor of thought, if not of humanity." According to Fumi Okiji, the project of Black poetics "involves what might be understood as outer-space exploration. It is a reconnaissance mission charting possible ways to approach the unthinkable—that is to say, ways to know blackness or simply ways to blackness."[99] Or in Jared Sexton's formulation, "Black life is not lived in the world that the world lives in, but it is lived underground, in outer space."[100] These are not mere metaphors. They are attempts to articulate something (almost) impossible to articulate.[101]

Epilogue

Over its course, this book has drifted ever further away from its stated topic: communication. But that movement is symptomatic, since, as we have seen, attempts to communicate with other species always—inevitably—end up somewhere else. Or to put it another way, communicating with another species is never simply about communication. By examining the more and less of communication, this book has also, at various times, veered more or less off its course.

Veering off into what lies beyond communication is constitutive of communication itself—this is a key argument of the book. To say that communication is never simply about communication is not some cop-out or excuse. The cop-out, if there is one, is to believe that one can study communication plain and simple, as though its excesses—love, desire, quasi-communicative systems, and so on—can be kept at bay through some kind of inoculation or autoimmune response.[1] Nor can race and gender be elided in a discussion of communication (whether intra- or inter-species), because they will return, as the repressed always does, inevitably and with a vengeance. The aim of this book has been to follow the excesses generated by the communicative endeavor, but also to continually return to the task at hand: to the subject of communication itself, our impossible subject. And so, in this brief epilogue, we come full circle.

Cops killed Elijah McClain in cold blood. McClain was not a scientist, but he felt deeply for lonely animals. The twenty-three-year-old Black man from Colorado spent his free time playing violin for cats at a shelter, in an attempt, he said, to soothe them. Lonely and abandoned, the cats were very likely soothed by his music. It was, in all probability, the young man's presence and empathy that mattered, not what style of music he chose to play, or even necessarily that he played music. To understand the interspecies encounter between McClain and the cats, then, it's not enough to focus on the notes. But the case of McClain also shows, as

Kodwo Eshun would say, that the human is a "treacherous category,"[2] for the young man was murdered by police while walking home from a convenience store. His only "crime" was being Black in an anti-Black world. The custodians of civil society—the treacherous domain of "Humans"—killed him for no reason at all.

In a recent interview, Frank B. Wilderson III provided a telescopic view of the last 1,300 years of human history, beginning with the argument that the advent of Blackness was coterminous with the advent of "social death," that is, existence outside the world of human social relations.[3] The major historical event in this regard was the establishment of transatlantic slavery, but anti-Blackness has earlier origins in Wilderson's account. He argues that the "Blackening" of African cultural identities—Ashanti, Igbo, and so on—produced Blackness as a "paradigmatic position, not a culture"

> that is produced through the Arabs, the Chinese, people who become Iranians and Iraqis, Moroccan Jews beginning in 625 AD. Somehow global consensus says, "Here's a place, we're going to call it Africa, and here are some beings—we're going to call them all Black, regardless of what languages they speak, or what anthropological accoutrement they have—we designate them as Black, and we say that they're socially dead. . . ." As the Arabs start developing who they are as Arabs, from 625 AD, and then the Portuguese take over. What begins to happen is the globalization of relations, that civil society starts to become a kind of global dynamic. And in order for there to be a civic entity, a civic realm, there has to be some libidinal, some psychic understanding, as to who is in and who is out. And war does not do that. War is a conflict between human agents. What is that outside? I think that Blackness has produced the inside-outside of a globalized civil society over the past 1,300 years.[4]

This is a devastating account, to be sure, and a controversial one at that. It emphasizes, above all else, just how treacherous the category of the human is. "There now exists a separate kind of human being, the American black man," Sun Ra said in 1969. "And I should say that he doesn't belong on this Earth."[5]

These quotations point to an increasingly widespread idea: genuine freedom for oppressed peoples would not require more humanity, but less—it would require the end (indeed, the abolition) of the Human *stricto sensu*. In Wilderson's view, overcoming anti-Black racism would mean the end of a number of categories we take for granted. Genuine liberation for Black people would not lead to a "Black existence that is whole." It would mean something totally different "on the other side." He says, in an interview with Aaron Robertson, "On the other side, it means something more catastrophic and renewing, which is having no Black existence because there will be no Black people. And having no Human existence, because there will be no Humans. There will be sentient beings who are on the cusp of a new episteme."[6] What would that mean for the environment, for our concepts of nature and of animals? Wilderson believes that we cannot possibly know.

A similarly pessimistic concept of humanity is increasingly evident in work on human/animal relations, and in particular human/animal love (the topic of chapter 2). But first, consider a more optimistic view. In an intriguing recent volume, Joanna Bourke holds out some hope for the possibility of good sexual relations between humans and animals. Like Margaret Grebowicz, Bourke sees this possibility in terms of queerness (after all, the history of prosecuting sexual "deviance" is thoroughly homophobic), but also turns to a different, and unlikely model: adult-infant communication.[7] To understand a possible, nonviolent love between humans and animals, she asks us to consider adult-infant communication, because in that case, too, genuine communion can take place *even without language* if adequate care is exercised. In Bourke's work, then, we see a familiar pattern in interspecies communication discourse: a kind of conceptual oscillation between communication and love, but this time in the other direction, for it is not an attempt at communication that leads to love, but rather an attempt at love which reaches—conceptually, at least—for communication as a model.

In response to Bourke's somewhat rosy suggestion, however, the philosopher Amia Srinivasan asks us to slow down and consider a less optimistic perspective.

> In sex of all things, where humans so often misconstrue what other humans want, where the temptation to project our fantasies and images onto the other is always pressing, where coercion and control figure all too easily—can we ever trust ourselves to know, really know, what an animal wants? Perhaps one day there will be members of the animal species *Homo sapiens* who are able to have sex with other animal species in a way that has nothing to do with the will to dominate, fetishise or transgress. If so, I think, those people would be of our species, but not of our kind.[8]

In a similar way to Wilderson, Srinivasan insists that exercising the kind of care necessary for a certain worldly encounter would require the end of humanity as we know it. The stakes of the argument are completely different, of course: Wilderson writes about Blackness and anti-Black racism, whereas Srinivasan addresses Bourke's work in the context of her interest in sex and consent. But the two converge on one point: certain forms of nonviolence will remain impossible so long as we remain human in the way we currently are.

In *The Symbolic Species,* Terrence Deacon writes, "Though we share the same earth with millions of kinds of living creatures, we also live in a world that no other species has access to."[9] He may be correct. But worlds have been constructed through means other than semiotic distinction, and access to those worlds has also often used different criteria. Beginning in the eighteenth century, Europe established precisely a world that excluded most people. The "Rights of Man and the Citizen" were applicable only to men. Hegel famously wrote that Africa is "outside history."[10] Transatlantic slavery turned Africans into commodities to be bought and sold. Black critical theory often emphasizes the singularity of transatlantic slavery. All other forms of slavery—from precolonial Africa to medieval Japan—have been based on conquest or punitive measures. But for the first and only time in history, transatlantic slavery was predicated on race, and on anti-Blackness in particular. Far from being contingent, Blackness was *ontologized.* This does not mean that Blackness *is* an ontological category but that chattel slavery turned it into one—not as an event that had been completed once and for all, but as an occurrence that had to be enacted repeatedly.[11] In short, the modern world was built on a series of exclusions that have little to do with the kind of distinctions that semiotic theories elaborate. This is why, for many (Black) theorists, genuine Black freedom would mean nothing less than the end of the world.[12]

But the world is already ending. We know this. Predictions about precisely when humans will become extinct range wildly, but it certainly seems that our days are numbered. For all the talk of higher-order cognitive functioning in humans—of technological, linguistic, musical, and artistic capacity—if humans destroy themselves (along with literally millions of other species), those marvelous attributes will not have amounted to very much at all. Although contemporary evolutionary theory has largely moved away from progressivism and teleology, the language of "higher" functioning is still typical, for example in the idea that symbolism emerges from an interaction of indexes and icons shifted to a "higher level." On its own terms, evolutionary discourse is justified; I am persuaded by evolutionary models presented by the likes of Deacon—as we all should be, I think. The hierarchical language deployed by biologists is the result of careful observation and analysis, not some kind of reflex value judgment.

If one thinks about what human capacity means for the vitality of our own species, things start to look different. For a while, humans might have been "winning" against their environment, but as Gregory Bateson famously put it, "The creature that wins against its environment destroys itself."[13] If we stay on the current path, humans will continue to kill millions of other species, until our own demise. If one adds to this forms of social violence like misogyny, racism, and homophobia—not to mention the genocides and famines of the twentieth century—the birth of an unprecedented form of misanthropy comes as no surprise. This new misanthropy would like to see humans gone sooner rather than later, so that we can do as little harm as possible in our twilight hour, move off the scene, and let other species flourish. The old idea that the end of humanity would be a great pity because of our tremendous "achievements" (art, beauty, society, and so on) seems less convincing when we are in the throes of a species-wide suicide, bringing harm to everything around us in the process.

Examples of this new misanthropy range from a popular T-shirt design that states, "Dogs: Because People Suck," to the Tiktaalik internet meme,[14] and the Voluntary Human Extinction Movement, which proposes that "phasing out the human race by voluntarily ceasing to breed will allow Earth's biosphere to return to good health."[15] The form of misanthropy I refer to is indeed unprecedented, as a recent and useful book by Andrew Gibson confirms.[16] According to Gibson, in its traditional formulation, misanthropy is characterized by its "impossibility" or fundamental "incoherence." As he argues in a reading of Molière's *Le Misanthrope* (1666), although the protagonist, Alceste, wants desperately to break with humanity, that "complete break is necessarily implausible or unachievable; true misanthropy is finally impossible."[17] A related feature of conventional misanthropy is that, ironically, it harbors a profound optimism, because it implies an expectation that humans should and can be better than they are.[18] Gibson does not make this point (at least not explicitly), but he suggests something similar when he observes that "part of Molière's admittedly complex thesis is that Alceste's misanthropy is actually in some degree a form of self-love."[19]

What I'm calling the "new misanthropy" breaks with past models in that it finally becomes coherent and possible.[20] The new misanthropy also jettisons the implicit optimism of conventional forms and takes on a more fundamentally fatalistic attitude to humanity. If there is anything "optimistic" about the new misanthropy, it is that the death of humanity would mean the flourishing of other, less destructive, forms of nonhuman life.

Abolishing the Human is the radical political horizon of much thinking about animals. But in the meantime, we continue to share the Earth with our co-species companions (and our co-species non-companions), our perceptual and semiotic worlds colliding, comingling, co-existing. How closely can those worlds be aligned? The question is far from resolved. And, unfortunately, when the question is asked, it is typically framed in terms of how far we can bring animals into our world.

The American psychologist Herbert S. Terrace once believed that at least certain non-human animals could be brought into the human world of language with a bit of training. In the 1970s, he worked closely with a chimpanzee named (ironically) Nim Chimpsky and managed to teach the ape sign language—or so he thought. In a widely publicized event, Terrace found (or believed that he had found) a fatal error in his own research: although it seemed that Nim understood certain words and could match them with signs, he was in fact responding to cues. "Instead of initiating a conversation, Nim first tried to reach for what he wanted (some crayons, a picture book, a banana, a pet cat, etc.). The teacher demanded that Nim sign by signing: *Where cat?*, *What color crayon?*, *More book?*, or simply, *Sign!* Having no alternative but to sign, Nim replied by imitating some of the teacher's signs and adding a few universally relevant signs such as *more*, *me*, *Nim*, and *hug*. Similar patterns of discourse may be seen in films of Washoe and Koko signing with their teachers."[21]

In 2019, Terrace wrote a book reflecting on the whole episode: *Why Chimpanzees Can't Use Language and Only Humans Can*. Others remain convinced that they have taught apes to use sign language, and the dispute (believe it or not) boils down essentially to whether apes are signing or whether researchers are imagining it. Part of the problem, it seems, is that those researchers who insist on the capacity for apes to sign are too close to the animals—are too intimately connected to them. The Gardners (Beatrix Tugendhut Gardner and Robert Allen Gardner) long advocated a "shift away from syntax . . . to the social and cognitive basis of language acquisition,"[22] claiming that such an approach allowed them to teach the chimpanzee Washoe well over a hundred signs. For them, in other words, communication is never just about communication. But this is also where the problem lies, because when you start saying things like that you lose scientific objectivity—this, at least, is the criticism leveled by Terrace and his proponents.

In a documentary about Nim, we see this dynamic play out clearly in a kind of cinematic dialogue between Stephanie Lafarge (the person Nim lived with while Terrace was trying to teach him sign language) and Laura-Ann Petitto (a Columbia University student and teacher at Project

Nim). Nim came to live at Lafarge's apartment in New York City, along with her husband and children, and Lafarge (like Margaret Howe with Peter) saw herself as a kind of mother to the chimp. The day Nim arrived at the apartment, recalls Jenny Lee (Stephanie Lafarge's daughter), "I think I fell in love instantly." And Lafarge remembers that Nim "liked alcohol. You'd give him a sip and he'd want more. We gave him puffs on a joint. We'd have to treat him like a child. We could expose him to the sensations we enjoyed." Petitto took umbrage with the way Nim was treated, saying, "I had an instinctive sense that something was very amiss here, that this is not the way you teach a child language. Or you interact with a child. Or you teach, um, anything language." Indeed, Lafarge seemed more interested in aspects of Nim's learning and exploration other than language. "When he began to discover my body, my nakedness, he'd be curious," she says. "I mean, like a child. He was uninterested, and then one day he was interested. I never felt sexually engaged with him. There was a sensuality but Nim was a . . . was a . . . pre-teen." The problem, thought Petitto, is that Lafarge was not interested in language at all. Because she was "a graduate student in psychoanalysis, her questions had to do with the Oedipal complex. And she was interested in Nim's masturbation. And how he masturbated. I couldn't believe it."[23] (The sexual vectors formed over the course of Project Nim are staggering: Terrace himself was involved with both Lafarge and Petitto, a dynamic that obviously contributed to the tense relationship between the two women.)

Disagreeing with the Gardners' position regarding the social basis of language acquisition smacks of scientific reductionism. Who, today, would advocate a kind of scientific "purity" in which the social is excised from language acquisition? Having said that, however, one can certainly overemphasize extra-communicative aspects, as when mother-child analogies are imposed on animals in a way that totally obfuscates animal behavior. Terrace was rankled by what he saw as the Gardners' moralistic judgments of his work (they referred to Nim's classroom as a "cell," for example).[24] He objected strongly to those criticisms, but also seemed to imply that, in emphasizing the "social" aspects of research, the Gardners had absolved themselves from scientific rigor and from tough questions surrounding the theory of language. In the end, Terrace caustically intones, the Gardners need to ask themselves what they are trying to discover. The title of his 1982 response to the Gardners' work is "Evidence for Sign Language in Apes: What the Ape Signed or How Well Was the Ape Loved?"

In brief, love *may* under certain circumstances be a useful tool in language training, but it cannot replace language training, which is the real object of study—this is Terrace's argument. In his final assessment, he argues that "before one studies 'the social and cognitive basis of language' one must, even at a minimal level, demonstrate a capacity for semantic and syntactic creativity."[25] In other words, you can't study the social basis of language in creatures without language (and he does not think that chimpanzees have language), and if you think you are doing so, you are probably deluding yourself. In allowing the question, "How well was the ape loved?" to supersede the question of ape sign language, Terrace thinks, the Gardners lost any ability to detect biases. Love may be blind, but what the Gardners experienced was worse than blindness: their love for Washoe made them see things that weren't there. It made them hallucinate, in a way.

This anxiety—that research on human/animal communication amounts to nothing more than a hallucination—has a precise starting point: a horse named *der kluge Hans* ("Clever Hans"). Beginning in the 1890s, Hans's trainer (Wilhelm von Osten) astounded audiences by showing that the horse could answer complex questions by tapping with his hooves or making other gestures. Hans appeared able to solve simple arithmetic sums, and identify colors and musical pitches. It turned out, however, that Hans was not so clever after all. It was revealed that von Osten was unwittingly tipping the horse off with minute head gestures.[26] Von Osten wasn't exactly a fraud—or perhaps he was, but mostly to himself.

The so-called "Clever Hans phenomenon" has haunted all subsequent research on the ability of animals to learn human languages. In 1980, nearly a century after the first appearance of *der kluge Hans* in front of audiences, Thomas Sebeok and Robert Rosenthal convened a symposium under the auspices of the New York Academy of Sciences titled "The Clever Hans Phenomenon: Communication with Horses, Whales, Apes and People." The motivation behind the event was largely to prove once and for all that such communication is not possible—to put the proverbial nail in the coffin, that is. Only a few scholars holding the dissenting view attended.[27]

At least according to Terrace, the whole scenario played itself out again with chimpanzees in the late twentieth century. A more balanced perspective is that, in this case, the evidence is equivocal. It would seem that little progress has been made over the century between Hans and us. Do chimpanzees deliberately sign with us? Or are we just imagining it? Worse than that, are we imagining it to such an extent that we unconsciously get chimps to do what we want them to?

With this larger context in mind, it's intriguing to reflect on what those elephants in Paris were actually doing when spectators reported, for example, that the song "Charmante Gabrielle" "plunged them into a sort of languor."[28] The powers of the imagination are strong.

But it's not only about animals that we have hallucinated. Today, we are easily deceived into believing that a chatbot (an algorithm, really) is a real person. As I write this, a well-publicized incident shows that the consciousness of AIs is a hotly contested topic—I am thinking of the Google software engineer Blake Lemoine, who was fired for claiming that a chatbot system called LaMDA became sentient.[29] In an excellent article about the affair, Ian Bogost writes, "I'm not going to entertain the possibility that LaMDA is sentient. (It isn't.) More important, and more interesting, is what it means that someone with such a deep understanding of the system would go so far off the rails in its defense, and that, in the resulting media frenzy, so many would entertain the prospect that Lemoine is right."[30] Bogost reminds us that there's a long history of exactly this kind of "going off the rails." As he notes, the MIT engineer Joseph Weizenbaum had already diagnosed the tendency in the 1960s. Reflecting on his research, Weizenbaum wrote in his book, *Computer Reason and Human Reason*, "What I had not realized, is that extremely short exposures to a relatively simple computer program [with nothing approximating sentience] could induce powerful delusional thinking in quite normal people."[31]

At the same time, one wonders what kind of communication is possible where desire and affection are completely absent. John Durham Peters levels exactly this charge against Alan Turing. Peters writes, "Turing's assault on consciousness as a guarantee of communication is admirable, but his veiling of love, attraction, eros, and morality is troubling. . . . What is missing from the Turing test—and the whole cultural complex that contemplates the replication of human beings in artificial intelligence—is the desire for the other that Hegel thought raised us out of animality into the homeland of consciousness."[32] We do not need Hegel (or his concept of the "homeland of consciousness") to make the point. Absence of desire for the other hollows out communication, while too much desire for the other tends toward delusion. Such is the dilemma of all communication, and not just the interspecies kind.

From a broad historical perspective, the Clever Hans incident preceded many examples in our own time, but it also had its own precedents. The eighteenth century is littered with examples of chicanery involving automata: machines that looked and acted so much like humans or animals that they could be passed off as real. Perhaps the most famous example is the so-called Mechanical Turk of 1769, a puppet-like contraption built in the form of a (highly Orientalized) Turkish man playing chess.[33] Its inventor convinced droves of audience members that the mechanical device could play a strong game, but it was revealed to be nothing more than a silly trick—for a human man was hidden within the machine. (It is after this eighteenth-century invention that Amazon's crowdsourcing website, Amazon's Mechanical Turk, is named.)

Visitors to fairs and exhibitions in the eighteenth and nineteenth centuries were aware of the skullduggery of show business, and it was in that context that P. T. Barnum famously entered the scene. In 1835, at a moment that one commentator calls "the birth of American popular culture," Barnum introduced spectators to Joice Heth—a slave who, according to accounts during the time, was previously owned by George Washington.[34] We know that Heth was captured in Africa as a child, and that she was a major boon for Barnum (who in his own words, built his vast entertainment empire on her display).[35] At a Boston exhibition, Barnum installed a version of the Mechanical Turk in a room next door to Heth, who was seated so that visitors could inspect her. She was, by this time, old and almost completely paralyzed. To drum up attention, Barnum penned an anonymous letter to the newspaper, which read,

> Joice Heth is not a human being. What purports to be a remarkably old woman is simply a curiously constructed automaton, made up of whalebone, India-rubber, and numberless springs ingeniously put together, and made to move at the slightest touch, according to the will of the operator. The exhibitor is a ventriloquist, and all the conversations apparently held with the ancient lady are purely imaginary, so far as she is concerned, for the answers and incidents purporting to be given and related by her are merely the ventriloquial voice of the exhibitor.[36]

Cultural theorist Louis Chude-Sokei notes the stunning irony in Barnum's description of "Joice Heth's" voice as "ventriloquial," since the author of this letter was almost certainly Barnum himself.[37] Barnum admitted to writing the following: the Mechanical Turk "somewhat prepared the way for this announcement [of Joice Heth's non-humanness], and hundreds who had not visited Joice Heth were now anxious to see the curious automaton; while many who had seen her were equally desirous of a second look, in order to determine whether or not they had been deceived."[38]

As Ellison would say, Joice Heth was invisible because (White) people refused to see her. "That invisibility to which I refer," wrote Ellison in *Invisible Man*, "occurs because of a peculiar disposition of the eyes of those with whom I come in contact. A matter of the construction of their *inner* eyes, those eyes with which they look through their physical eyes upon reality."[39] So much of what I have written about in this book is attributable to inner eyes and inner ears—to the projections of those unable to see and hear beyond themselves.

For those who believe that music is a more flexible and emotional form of communication than language, interspecies improvisation offers an attractive opportunity. But as I wrote in this book's introduction, there, too, people seem able and willing to fool themselves into a substantial encounter: recall Peter Gabriel playing music with a chimp who tinkers with white keys on a piano. David Rothenberg has been at the interspecies improvisation game for a lot longer than Gabriel, although the efficacy of his work also remains equivocal. How can he ever be sure that he has broken through the interspecies barrier (as John Lilly would put it)? He admits this uncertainty in the wistful conclusion to his book, *Thousand Mile Song*, recounting a question put to him by Paul Knapp, a fellow whale enthusiast. "'Did you manage to do it, did you ever get through to a humpback whale?' He hands me a cup of hot tea. The wind rustles in from the ocean through the palms, blending its sound with the bleats of forest frogs. I gaze out toward the mouth of the bay, remembering all I've heard. 'The more music I make with whales, the less I know. Maybe he listened to me, Paul, or maybe not. But have I got a song for you . . .'"[40]

✷

One of the most prolific historians of the extraterrestrial life debate recently wrote a book arguing that, in light of recent research on exoplanets, the discovery of extraterrestrial life is likely (rather than merely possible). He argues, therefore, that we should start formulating policy right now to prepare for the social impact that such a discovery will certainly have.[41] In the long (if not necessarily venerable) tradition of books such as these, I conclude by considering the hypothetical scenario of humans encountering extraterrestrial *intelligent* life. I do not have anything mind-blowing to contribute, but based on a decade of solid research, I will simply say this: should an extraterrestrial spacecraft land on our planet, could we please not send Avi Loeb, Robin Hanson, or Elon Musk as part of our greeting party? And no Herbert Terrace please, nor any US president.

Because communication is always about more than communication, let us send a group of humans whose intellectual, linguistic, ethical, and political commitments are up for the task. Let us send Alexis Pauline

Gumbs, who knows how to listen, and who knows how to love. Let us put our best foot forward, allowing Ted Chiang to join the greeting party—a creative, generous person. Gary Tomlinson would help the group think things through in a sufficiently complex manner, and Kapwani Kiwanga would know just the right pose to strike, just the right word to use and when. Amia Srinivasan would keep everyone grounded, and Frank Wilderson would compel us to look unflinchingly at everything (in ourselves and in the alien) we may want to avoid. The spirit of Sun Ra (if I may be permitted one further hyperbolic gesture) would set the mood just right. (I may also suggest taking along a couple of other individuals who I have not had an opportunity to discuss in this book: Don Mee Choi, for her translational acuity and Stanley Jordan, for his hawk ears.)

What happens next is anyone's guess.[42] But by approaching the interspecies encounter as a matter of communication saturated with ethics and politics, humanity might have hope. Or, even more optimistically, we may end up finally becoming something radically other than human.

✷

But that is not the end, because this book is not primarily concerned with a horizon where the human, having spent its humanity, lapses into something inhuman. It focuses, instead, on attempts to trespass human boundaries, to enter into communion with beings unlike ourselves (whoever "we" are).

In the last couple of years, writing about the perceptual and communicative particularities of various animal species has mushroomed. As one author put it, each animal "is enclosed within its own unique sensory bubble, perceiving but a tiny sliver of an immense world."[43] This is laudable work, and the message (namely, that we should think twice before anthropomorphizing) is well taken. But surely that cannot be the end of the story, because all of us—the sniffing dog, the language-using human, and the echolocating dolphin—share the Earth. What's more, the enclosed bubbles of animal perceptual systems encounter each other, they meet. Those encounters are the topic of this book. Because of the overlapping region between human and non-human signs (humans, birds, dogs, and dolphins are all experts at using indexes, for example), interspecies communication is not only possible but inevitable. I like Eduardo Kohn's term for communication that happens in these overlapping regions: transspecies pidgins.[44]

I'm not going to end this book by saying that we don't need to gaze up into the night sky because the world around us is full of chatter. Nor am

I going to celebrate the sensory realities of ticks, bats, and dolphins. We need less celebration and more careful thinking, less nostalgia and more clear-eyed historical awareness. In fact, as I continue work on this topic, I feel myself drawn increasingly toward simple questions. Aside from the bigger political quandaries surrounding humanism and animality, making headway on the subject of interspecies communication would be greatly aided by some fairly basic knowledge.

For one thing, it would be interesting to know why whales vocalize the way they do—that is to say, why they "sing." It would even be worthwhile spending some time thinking about how to pose the question correctly, since the "why" of animal behavior is no simple matter. Knowledge of what motivates whales to vocalize in elaborate patterns would greatly help our understanding of what whale vocalization is. A satisfying answer would require working through a lot of terms (as I did to some extent in chapter 1), while also warding off simplistic explanations like those offered by Roger Payne.

It would also be nice to know whether apes and parrots have the capacity for symbolic meaning or whether we are simply imagining it! The fact that I had to write the previous sentence indicates the graveness of the problem. But surely with the requisite self-reflexivity there's a way to find out? Surely we are not doomed to forever wander in the radical uncertainty of never knowing whether an ape is signing or whether people are simply deluding themselves? This is a question with an answer, at least to some extent.

Coming up with ideas about how best to cope with the cosmic predicament of not knowing whether "we are alone" would also be helpful. Rather than assuming that we will one day find out, or putting all our eggs in the Avi Loeb/Elon Musk basket, might we not instead accept (tentatively, at least) that we may never have a satisfying answer to the question? Why not accept, in other words, that we don't know whether we will ever know? What would *that* mean for our lives on planet Earth today? If, on the one side, we probably can find out whether apes are capable of sign language and therefore have no reason to resign ourselves to radical uncertainty on the topic, on the other side, we perhaps should "resign" ourselves (at least until we know more) to radical uncertainty on the topic of extraterrestrials.

In the final analysis, interspecies communication may simply mean learning to live with human incomparability, while simultaneously exploring the regions of meaning we share with other creatures on earth. There are no shortcuts in either endeavor, only the slow and unsentimental plodding of thought. That, in any event, is what I have attempted here.

Appendix

A SUMMARY EXCURSION INTO BIOSEMIOTICS

In this appendix, I sketch the outlines of a particular biosemiotic theory, but my aim is not to present anything like a comprehensive picture, which would require a book in itself. Nor do I pretend that researchers have worked out all the nuances of this theory, or even that it is completely correct. Revisions are inevitable, and researchers will be working on the topic for many years to come. (What's more, there are some versions of biosemiotics from which I take considerable distance—hence the insistence on the particular version I lay out here.)[1]

None of that should detract from the insights that have been garnered, however, nor should it allow us to regard the whole endeavor as a social construction. At a time of climate change denial and anti-vaxxer rhetoric, a commitment to the realist ontology of scientific domains is urgent. In other words, biosemiotics is not metaphysics. It is an attempt, however imperfect, to grapple with facets of animal behavior outside the realm of human intelligence and symbolic forms of communication. All I hope to do here is set some basic coordinates.

We turn to C. S. Peirce, who famously developed a tripartite model of the sign: iconic (based on resemblance), indexical (based on proximity or causal association), and symbolic (based on convention).[2] In all three cases, more is required than a mere relation between sign and referent. All semiosis requires a fourth aspect: namely, something capable of "interpreting" signs.[3] Interpretation here is taken in the broadest possible sense as the relation *to* a sign.[4] Attempting to define this metarelational structure, Peirce called the translation of the sign/object relation by an organism the interpretant.

Unlike the other two types, *symbols* are the domain of humans only.[5] Symbols have several features: they bear a primarily conventional relationship to their referents; they constitute self-sustaining systems of meaning that create a gap between the sign system and the external world; this gap means that symbols can be employed at a far remove from a designated referent (e.g., one can use symbols to describe a faraway object, as well, of course, as an abstract idea); and because of the first three features, symbols require a larger system of grammar or governance in order to function properly.

Indexicality, by contrast, requires only the capacity of an organism to learn from past events and apply knowledge garnered from experience to new situations. For example, because smoke is a by-product of fire, many animal species will register danger already when seeing or smelling smoke and before any direct encounter with a fire.[6] In addition to humans, most mammals and species of birds use indexes.

Icons function merely by means of similarity, that is, by an animal's ability to recognize that something is like or unlike something else.[7] "In bringing the presence or absence of difference to the heart of semiosis," writes Gary Tomlinson, "iconicity can be seen to underlie systems of innate animal response to the world, whereby aspects of external stimuli are taken to identify food, predator, or mate."[8]

Symbols, indexes, and icons are all signs. It is important to distinguish between signs and what we might usefully call "signals," where a signal, unlike a sign, requires no interpretation (in Peirce's sense) and, therefore, no meta-relationality. Bacteria, to use the most typical example, communicate in the sense that they send signals to each other. It makes sense to say that bacteria communicate, provided that we understand communication as a form of information transmission—as mere "correspondence in the face of randomness"—and not as a series of content-laden signs.[9] In this way, the human, dog, whale, or bird as a sign-user is of a different order from the bacteria as a signal sender/receiver. Even so, we are interested in gradations, not categorical breaks, and there are certainly instances where signs are only minimally separated from signals. There are also disagreements about where information processing (through signals) ends and where semiosis (through signs) begins. In some accounts—for example in the work of Kohn—the tick is held up as the icon user *par excellence*. Tomlinson, by contrast, sees no meta-relationality in the tick and would thus reject Kohn's argument.

Returning to Peirce's model, we see that his three kinds of sign are functionally related. Indexicality is dependent on what I would call temporalized iconicity, which is another way of saying *repetition*. The capacity to associate smoke with fire (indexicality) means nothing in itself if an organism is unable to recognize smoke again the next day. To re-recognize smoke—to remember it—requires a kind of iconic semiosis capable of detecting similarity across time.[10] This is why indexical semiosis is only possible in creatures with significant memory and learning resources. The capacity to remember and learn "enables some species to kick iconic perception onto a higher level, yielding indexical association. Indexicality is iconism transcended and thereby transformed."[11] By the same token, symbolism employs the resources of both iconicity and indexicality but pushes them to a new register.[12] Indexes begin pointing to one another, rather than to external objects, and begin circulating in a semi-autonomous system of meaning. Although indexicality relies on iconicity, and symbolism relies on both, the emergence of one kind of sign system from another necessitates more than addition or accumulation. The emergence of symbolism is exactly that: *emergent*, meaning that symbols rely on the interaction between indexes and icons but are not reducible to them.[13]

From the vantage of biosemiotic theory, the distance between different forms of animal existence begins to narrow. Whether one emphasizes the differences or similarities between symbolism and other forms of semiosis depends largely on what one hopes to achieve. The symbolic meaning of human language is both intimately related to—and of a different order from—other forms of semiosis.

In his ethnography of an Amazonian forest community, Eduardo Kohn invites the reader to think about semiotics expansively—encompassing humans, jaguars, and peccaries—without making the mistake of ventriloquizing non-humans. He does this by provincializing language (that is, the arbitrary signs of symbolism) that many have mistakenly taken to count for semiosis all told.[14] Kohn argues for a broader sense of semiosis capable of accounting for the specificity of human forms of communication, while simultaneously investigating how the human semiotic practice "grows out of and is constantly in interplay with" the world of broadly distributed biosemiosis.[15]

Symbolic meaning is the hallmark of human communication, of course. And as I have mentioned, the symbolic system is quasi-autonomous. But consider two other forms of human expression: non-symbolic "words," and music. Narrating to Kohn the story of how he shot a collared peccary in the forest, an Ecuadorian hunter named Maxi exclaimed: "*tsupu*." Kohn leaves the word deliberately untranslated, asking the reader, "What might it mean? What does it sound like?"[16] As it turns out, the word refers to an object tumbling into a pool of water—just like Maxi's wounded peccary, which ran headlong into a river. In contrast to symbols, however, *tsupu* is not merely conventional but rather resembles the action to which it refers. (In colloquial parlance we would call it an onomatopoeia.) The word *tsupu* (but is it a *word*? Kohn asks) begins to point outside the enclosure of symbolic language.[17] It alludes to a practice of language that is open to the outside world by virtue of a mechanism that is not entirely conventional.[18]

For some, a "word" like *tsupu* may intuitively seem like the becoming musical of language. (For reasons I will explain, this is an idea that I do not accept.) Consider, for example, an assertion by well-known musician and ecologist David Rothenberg: "Through music you can cross cultural lines, not because music is a universal language, but because it is a fluid and emotional form of communication, able to embrace and make use of unknown sounds. Something new can be produced between people from different sides of the world, *or even between different species*."[19] Rothenberg takes music's "fluidity" as a characteristic that might help open communication between cultures or even between species. The implication is that language is the exact opposite of music, in that the former *creates* rather than dissolves boundaries. (Rothenberg: "You make music together with those with whom you cannot speak."[20] This, I think, is an indefensible position.) Nor is Rothenberg alone in this kind of assertion. The idea that music is more primordial, more emotive, less rational, and less bounded than language can be found in Rousseau and Darwin and all the way to Carl Sagan and contemporary popular writers like Steven Mithen.[21]

It is a mistake, however, to understand music as emotional, prelinguistic expression. It is equally inadequate to define music as having *more* fluidity or *less* semantic specificity than language.[22] All such negative definitions lack clarity. To develop a positive definition of music, it helps to think about semiosis in evolutionary terms. Based on the available evidence, the most plausible hypothesis regarding the evolution of music and language is that they rely equally on complex cognitive capacities, which they harness in different ways. In other words, music is not a superficial by-product of language, nor is language a rational formalization of "musical" expression. (Both positions—erroneous in my view—have advocates.) Instead, music and language can be viewed as two independent but overlapping systems with complex historical associations. To some extent (although this tells only part of the story), human music and language find their evolutionary basis in *gesture-calls*, those partly involuntary and largely emotive expressions shared widely across mammals: think of a dog's snarl, or a vervet monkey's warning call, or human gestures like sighing and laughter.[23] Gesture-calls are always indexical, pointing either to an external referent (an intruder, a falling branch, a desired foodstuff) or an inward state (fear, hunger).

A defining characteristic of gesture-calls is that they lie on a continuous spectrum: in the canine realm, from a short and ominous snarl to a loud and sustained growl; in the human realm, "from a giggle to a laugh to a guffaw to a snort to a cry, for example, or from a slight to an exaggerated shrug."[24] Gesture-calls are, from this perspective, self-sufficient expressive units: they cannot be broken down into discrete component parts and reassembled to mean something else. A dog cannot combine the first part of a snarl with the end of a bark to mean something altogether new. Humans would only be able to combine a fragment of laughter with a shoulder shrug in an abstract, non-indexical manner.

I repeat: gesture-calls cannot be broken down into discrete component parts and reassembled to mean something else. However, at some point in our long evolutionary history hominins did begin to pull gesture-calls apart, thereby transforming them into something else entirely. Early hominins must at some point have begun to break down and reassemble gesture-calls in ways that released them from their original functioning. This release, or "winnowing" as Tomlinson calls it, ushered in something of an evolutionary transformation in which, for humans, gesture-calls were rendered something else entirely. To put a sharper point on it, the release of gesture-calls from their original function inaugurated a new type of sign.

One trajectory of this novel sign category resulted in what would finally become the phoneme: in English, "ges" and "ture" are not signs at all, but they become signs when placed in specific combinations, such as "gesture," "gestation," or "pasture." But a diverging evolutionary line led to discrete pitch, a category that, with Tomlinson, I take as fundamental to any biosemiotic theory of music. Unlike gesture-calls, which are analogue in the sense of lying on various spectra (loudness, length, and so on), musical pitch systems are by and large made up of sets of discernable, individual notes. A distinguishing feature of human cognitive capacities, then, is the deployment of discrete elements in various combinations or patterns. But the varied sign systems humans use differ considerably when it comes to meaning or reference.

In the case of language, "meaning was strengthened by the combinations of . . . phonemic units into words."[25] And words, when used in context, are wonderfully (if sometimes ambiguously) referential, whether they refer to things in the world (objects), abstract concepts (time, negation), or other things. Musical pitches, by contrast, were never again so closely linked to meaning. To this day they carry minimal and vague indexical associations, acting as signs primarily in contexts that also involve modern symbolism. This is a strong claim, and I invite those skeptical of it to continue reading below. What I want to emphasize now, in a slightly overstated way, is that the abstracting of pitch from meaning represents a momentous swerve in communicative means as, for the first time in the long development of hominin communication, a new ingredient appeared in vocalized gesture that attenuated meaning and referentiality rather than bolstering and specifying them.[26]

Music's referentiality is a controversial topic.[27] And indeed, one can think of many examples around the world where rhythms or melodies reference things: the major mode in Western classical music might reference happiness, the theme from *Jaws* has an ominous ring to it, and the tendency to impose narratives on musical passages has been widely documented (when asked directly, amateur listeners will often come up with similes such as "it sounds like a cat chasing a mouse," or "it sounds like friends playing in the sunshine"). The extent of music's referentiality also varies widely across cultures. While in some (probably most) cultures music typically references an affective quality or mood, instances exist where music hews more closely to symbolic meaning, such as the famous case of African talking drums (which, because they work in conjunction with tone languages, can convey fragments of properly linguistic meaning). Even so, it is difficult to disagree with the general outlines of Kofi Agawu's differentiation between language and music: "There remains a categorical difference between a system in which the word 'kettle' harbors a core reference, . . . and one whose 'vocabulary,' itself already contested and elusive, remains diffuse, amorphous, and devoid of a priori meaning."[28] Music usually functions in the "poetic" (Agawu's term) realm, while language usually functions outside the poetic realm. Although he does not define the poetic directly, we have a sense of what Agawu means by the term when he argues that, in a reverse tendency to language, music uses features like repetition that thrive on "informational redundancy."[29] Needless to say, people use language in manifold dialects and on a plethora of poetic registers; what Agawu has in mind (as I do) is linguistic tendencies. (Saying that language usually functions outside the poetic realm implies that language does, of course, sometimes function within the poetic realm—as poetry, for example. And poetry, like music, often thrives on informational redundancy.)

Returning to the evolutionary model and to the brachiation of language and music from gesture-calls, Tomlinson offers a pithy definition of music as the "extended, spectacularly formalized, and complexly perceived systematization" of non-symbolic, discrete pitches.[30] Language is a similarly extended, formalized, and complexly perceived systematization of symbolic signs (or, be more precise, the formalization of both phonemes and of symbolic signs, or words). But the kinds of extension and formalization are different in the two systems, to say nothing of the strikingly different results.

Which brings us back, finally, to the peccary hurtling into the river. Having trudged through some dense biosemiotic terrain, we conclude that music functions in exactly the opposite way from expressions like *tsupu*. Whereas *tsupu* is a word trying to escape the enclosure of language, music is non-symbolic meaning bound up in its own self-sustaining system. In evolutionary terms, one might say that music and language are two "examples of sonic materials that . . . have for some reason separated themselves from genetic matter and taken flight, following their own evolutionary line."[31]

As self-sustaining systems, language and music have a certain level of autonomy—at least compared with indexical signs (i.e., indexes), which point directly to their referents. Language is partially, but not completely, autonomous: symbols obviously reference things outside themselves. The mechanism by which language does this is bafflingly complex. In Peircean terms an approximation would go something like this: The mediating gap that symbolism famously introduces into semiosis is "filled in, again" by a "second order of indexicality operating within symbolism."[32] Crucially, this second-order indexicality does not happen at the level of the individual symbol; it requires mobilizing the entire symbolic apparatus. For example, whereas smoke directly indexes fire, using symbolic language to refer to a particular fire requires a shared knowledge of linguistic conventions (e.g., the English word *fire* or the Zulu word *umlilo*), as well as a grammatical structure that would allow one to combine symbols into a larger organization (e.g., "the fire that is in front of us now"). This is what Terrence Deacon means when says that "we can positively define symbolic reference as reference mediated by a closed system of indexical relations which, taken together, refer holistically to a system of relations in the world."[33] That symbols do not have a direct relationship to referents (there is no inherent association between word and object) has another important consequence: the meaning of a symbol depends on the context of its use as well as on its combination with other symbols. The fire in front of me now is distinct from the fire in my heart.

Music presents an even more serious challenge to referentiality. As I have mentioned, within the broadly (bio)semiotic terrain I am operating in here, emphasis must be placed on non-symbolic discrete pitch as a fundamental unit that is systematized according to various structures. Pitches are systematized by bringing them under the governance of tonal and other pitch-based structures. But in music, pitches are also brought into relation with structures that are themselves pitch-independent. For example, pitch organization has a temporal dimension, which brings parameters like duration, rhythm, and meter into the picture. In Tomlinson's account, properly human music making is characterized by the coalescence of (1) discrete pitch combinatoriality, and (2) temporal synchronization through entrainment.[34] The durational or rhythmic side of the story requires a fuller elaboration than I have provided here.

While musical figures do sometimes assume indexical meaning, the autonomous (or semi-autonomous) functioning of musical systems also makes them especially prone to metaphysical imaginaries.[35] To state things in slightly different terms, because music often functions as a non-symbolic, quasi-autonomous sign system, it is frequently registered as a kind of fantasy.[36] Music's strange mixture of spectacularly organized non-symbolic pitches propelled by temporal entrainment and limitless timbral variety makes referentiality hazardous, to say the least.[37] Music's relationship to referents—whether metaphysical or proximate—is thoroughly underdetermined. And indeed, music "thrives on that ambiguity."[38]

What people often refer to as "animal song" (or "animal music") is an in-between case. Neither birds nor whales (the two archetypal "musical" animals) deploy discrete pitches in their "songs." Of course, they don't deploy anything like phonemes for communication either. Instead, birds and whales both use gesture-calls by abstracting them and then ordering them in regular patterns. Bird song and whale song as abstracted, patterned gesture-calls: this definition is, if I may be so bold as to say so, accurate and elegant. It's astounding to recognize, then, that I am the first to provide it. Although many people have said many interesting things about bird and whale vocalization, they almost inevitably get tripped up in one of two places: (1) they imprecisely and wrongly understand music as a more emotive or looser version of language (passing over, that is, the fine distinctions between different kinds of signs that constitute both); and (2) they mix and match terms in unscientific and ultimately unhelpful ways.[39]

So where does this leave us? To be sure, the preceding excursion into biosemiotics is neither perfect nor complete. In my discussion of language's and music's building blocks (phonemes in language, pitches in music) I skirted various hybrid cases, such as languages that use tones (that is, tone languages), or the linguistic features of certain musical practices.[40] These counterexamples will not damage biosemiotics theory as much as they will help to nuance it. I, like the writers I have drawn on in this section, welcome elaborations, refinements, and, where necessary, serious revisions.

I would like to add one further point specifically about musical semiotics. As I've already noted, animal behavior researchers resort to music when they come up against something they do not fully understand. Now, in some sense, they do this in an unscientific way, appealing to music as kind of DIY suture when other, more easily comprehensible (and analyzable) concepts fail. They do this through what I would call a metaphysicalization of music: music no longer as a sign system but rather as something metaphysical that can be used flexibly to paper over intractable problems. (As mentioned, thinking of music this way has old roots.)

When equipped with carefully assembled biosemiotic tools, however, it is possible to analyze music in a non-metaphysical fashion. But this is not the end of the story, because, as I have suggested, to "understand" music as a semiotic system is precisely to understand the ways that music produces metaphysical imaginaries. In this sense, metaphysics is an effect deliberately produced by music (and by musicians). The skilled performer who spends hours practicing each day knows that music requires knowledge, discipline, and skill—hardly the hallmarks of pure emotion and affect. At the same time, most musicians would agree that the aim of their performance is not to somehow convey how much knowledge they have acquired, how disciplined they've been. If the system and the process of making music is always, at least to some extent, scientific (and here I use that word in all its range of meanings), the result is more often emotional, affective, or richly imbued with imagination.

Such, I would argue, is music's central paradox. As one author has influentially argued, ineffability is central to (and constitutive of) music.[41] Biosemiotics helps us understand why this is the case. It allows us to explain why music is ineffable without, however, explaining that ineffability away. Biosemiotics helps us explain how and why music produces something that exceeds explanation.

Music, in this sense, is both knowable and unknowable. I argue that non-human animals are also knowable to some extent. There is no reason to accept Nagel's strong claim that we can never know what it is like to be another animal, such as a bat. As Eduardo Kohn says, "We can know something of how red might be experienced by a blind person," or "what it might like to be a bat, . . . however mediated, provisional, fallible, and tenuous these understandings may be."[42] At the same time, we can never know everything it means to be a bat. A certain threshold of unknowability shapes—one might even say haunts—human/animal interaction, and in that space beyond the known and the unknown, many things happen. The known, the unknown, and the unknowable have been key terms in this book.

Peirce had his own take on the subject. In a text from 1885, he wrote that all questions "that are either practical or could conceivably become so are susceptible of receiving final solutions," because, even if humans cannot contrive the solutions, "we may take it as certain that other intellectual races exist on other planets,—if not of our solar system, then of others; and also that innumerable new intellectual races have yet to be developed; so that on the whole, it may be regarded as most certain that intellectual life in the universe will never finally cease."[43] In principle, then, at least according to Peirce, a comprehensive knowledge of animal semiotics is possible, if not by us, then by our fellow animal thinkers on other, future planets.

Acknowledgments

This book began during a residency at Akademie Schloss Solitude in Stuttgart. There, I would like to thank Angela Butterstein, Anita Carey-Yard, Chandrahas Choudhury, Thomas Crowley, Luzi Gross, Rasha Hilwi, Dima Hourani, Janine Jembere, Jean-Baptiste Joly, Imran Ali Khan, Hasan Mashni, Shaka McGlotten, Jasmin Schädler, Lotte Thieroff, Kostis Velonis, and Elisabeth Weydt. To bookmark the process, as it were, this project was completed during a Humboldt Foundation Fellowship. In Berlin, I owe thanks to Sebastian Däschle, David Frühauf, Katharina Hartwell, Dagmar Schäfer, K. P. Singh, Hanna Stief, and Viktoria Tkaczyk.

Also along the writerly way: I thank my editor, Elizabeth Branch Dyson, for her tireless support. Mollie McFee was amazing through the process, and I'm grateful to Nicholas Murray for his sharp copyediting. Caterina MacLean once again did a stellar job shepherding the book through production. Emily Murdoch Baker and Roy Sablosky provided additional editorial wisdom.

At Princeton, a number of folks have actively helped with crafting this book. Thanks go to D. Graham Burnett, Steven Chung, Asif Ghazanfar, Simon Gikandi, Wendy Heller, Harvey Lederman, Tom Levin, Lisa Margulis, Simon Morrison, Beth Schupsky, Juri Seo, Jeff Snyder, Emily Thompson, Daniel Trueman, Dmitri Tymoczko, and Autumn Womack. Now would be a good moment to add that this publication is made possible in part by support from the Barr Ferree Foundation Fund for Publications, Department of Art and Archaeology, Princeton University. Tim Noakes at Stanford Libraries was also very helpful.

To those I've worked with, written about, and learned from: thank you! Especially Kapwani Kiwanga, Sizwe Miya (may you rest in peace, *mfowethu*), and Wilson Salukazana.

Much gratitude to friends in the intellectual community for providing a platform for my work: Marié Abe, Michael Birenbaum Quintero, Jocelyne Guilbault, David Gutkin, Rujing Huang, and Michael Weinstein-Reiman.

To the friends, mentors, and teachers who have discussed the book with me: I appreciate you. I am thinking of Kofi Agawu, Manuel DeLanda, Bill Dietz, Brent Hayes Edwards, Chris Frank, Michael Gallope, Roger Grant, Bob Labaree, Ian MacMillen, Nicholas Mathew, Peter McMurray, Jairo Moreno, Ana María Ochoa Gautier, Benjamin Piekut, Ronald Radano, Anthony Reed, Anica Rissi, Kirsten Saracini, Gary Tomlinson, Hong-Kai Wang, and Michaela Wünsch. And a special thanks to Roger Grant and Helen Kim for organizing the South Philadelphia chapter of the Order of the Dolphin symposium.

My family has been generous and patient. Thanks to my father, Len, for always asking how it's going, to Brett for arguing with me, and to Joel for help along the way. My mother, Sally, was an amazing interlocutor until her final days. I was and am very fortunate.

She gets the dedication page, of course, but what would I be without my everything person, Helen Hwayeon Kim? The answer, of course, is nothing at all. Words fail me. I have only food, music, and an infinite, (in)human love to offer.

Notes

Preface

1. Michael Schetsche, René Gründer, Gerhard Mayer, and Ina Schmied-Knittel, "Der maximal Fremde: Überlegungen ze einer transhumanen Handlungstheorie," *Berliner Journal für Soziologie* 3 (2009): 469–91. The authors provide the examples of dolphins (in a therapeutic context), extraterrestrials (in the context of "search for extraterrestrial intelligence" programs), and demons (in the context of Catholic exorcism). These "strangers" are all very different from humans, but each exhibits characteristics that make it a potential, if uncertain *Interaktionspartner* (interaction partner, 272).

2. Kant typically addresses this tendency toward error under the banner of "Paralogism." See *Kritik der reinen Vernunft* (1781).

3. I use the term in the sense famously developed by Jacques Derrida in, among other places, *Of Grammatology*, trans. Gayatri Chakravorty Spivak (Baltimore: Johns Hopkins University Press, 1976 [1967]).

4. John Durham Peters, *Speaking into the Air: A History of the Idea of Communication* (Chicago: University of Chicago Press, 1999), 21.

5. Peters, *Speaking into the Air*, 30.

6. Peters, *Speaking into the Air*, 112.

7. Peters, *Speaking into the Air*, 244.

8. The Wikipedia entry on "interspecies communication," for example, has links to nine pages under "See also," one of which is the page on Lilly. Other suggested entries are "Animal communication" and "Great ape language."

9. An early reader of this book said that the colonial dimension of these authors is the least interesting aspect of their work. I would phrase it a little differently, however. If I were *only* saying that certain authors were stuck in a colonial mentality, that would be unsatisfying and not very useful. But what I do, instead, is begin with the observation that colonialism and racism shape much of the extraterrestrial life debates and go from there. It's a beginning observation, in other words, not the slam of an anvil.

Introduction

1. The authors make their intervention in the context of a thorny and specific academic context. Diagnosing the convergence of musicology and New Historicism in the 1990s, they develop the notion of quirk historicism. They write, "We can see

in this microhistory of musicological microhistories something of the iterative tendency of the historical quirk, adapting as it does to disparate discursive contexts. [It is clear] how much sheer pleasure can be had from researching and writing in the quirk historicist vein. There is the initial thrill of encountering the unexpected thing, the fluidity with which it is possible to add new links to the chain, and the quick payoff: each detail is so new and surprising as to seem to transmute instantly into historical gold." See Nicholas Mathew and Maryann Smart, "Elephants in the Music Room: The Future of Quirk Historicism," *Representations* 132.1 (2015): 61–78, 63. They refer also to James Johnson, *Listening in Paris* (Berkeley: University of California Press, 1994), 129–30. On the conjuncture of music and sex, see Jeffrey Kallberg, "Peeping at Pachyderms: Convergences of Sex and Music in France Around 1800," in *Fashions and Legacies of Nineteenth-Century Italian Opera*, ed. Roberta Montemorra Marvin and Hilary Poriss (Cambridge: Cambridge University Press, 2010), 132–51. See also John Deathridge, "Elements of Disorder: Appealing Beethoven vs. Rossini," in *The Invention of Beethoven and Rossini*, ed. Nicholas Mathew and Benjamin Walton (Cambridge: Cambridge University Press, 2013), 305–32. For an interesting take resembling my own, see Arman Schwartz, "Don't Choose the Nightingale: Timbre, Index, and Birdsong in Respighi's *Pini di Roma*," in *The Oxford Handbook of Timbre*, ed. Emily Dolan and Alexander Rehding (New York: Oxford, 2018), 433–63, 435.

2. David Sulzer, *Music, Math, and Mind: The Physics and Neuroscience of Music* (New York: Columbia University Press, 2021), 239. See also R. S. Heffner and H. E. Heffner, "Hearing in the Elephant (*Elephas maximus*): Absolute Sensitivity, Frequency Discrimination, and Sound Localization," *Journal of Comparative and Physiological Psychology* 96.6 (1982): 926–44.

3. Sulzer, *Music, Math, and Mind*, 239.

4. Clarence J. Glacken, *Traces on the Rhodian Shore: Nature and Culture in Western Thought from Ancient Times to the Eighteenth Century* (Berkeley: University of California Press. 1976).

5. Natural selection was unknown until the work of Darwin and others in the nineteenth century; in the eighteenth century Lamarckist accounts of evolution were dominant.

6. M. de Fontenelle (Bernard Le Bovier), *Conversations on the Plurality of Worlds*, trans. Elizabeth Gunning (London: J. Cundee, 1803 [1686]), 64.

7. Fontenelle, *Conversations*, 85.

8. Fontenelle, *Conversations*, 51.

9. Metta Bryld and Nina Lykke, *Cosmodolphins: Feminist Cultural Studies of Technology, Animals, and the Sacred* (London: Zed Books, 2000), chap. 7 ("Voices from Inner and Outer Space").

10. John Lilly, *Man and Dolphin: Adventures on a New Scientific Frontier* (New York: Doubleday, 1961), 217.

11. On synchronization of movement, see Charlotte Duranton, Thierry Bedossa, and Florence Gaunet, "Interspecific Behavioural Synchronization: Dogs Exhibit Locomotor Synchrony with Humans," *Scientific Reports* 1 (2017):12384. On dog responses to the human voice, see Ikuma Adachi, Hiroko Kuwahata, and Kazuo Fujita, "Dogs Recall Their Owner's Face Upon Hearing the Owner's Voice," *Animal Cognition* 10 (2007): 17–21. On contagious yawning, see Ramiro Joly-Mascheroni, Atsushi Senju, and Alex Shepherd, "Dogs Catch Human Yawns," *Biology Letters* 4.5 (2008): 446–48. But see also A. L. Harr, V. R. Gilbert, and K. A. Phillips, "Do Dogs (*Canis*

familiaris) Show Contagious Yawning?" *Animal Cognition* 12 (2009): 833–37. On shepherd calls, see Christian Kaden, *Hirtensignale: Musikalische Syntax und Kommunikative Praxis* (Leipzig: Deutsche Verlag für Musik, 1977). For an outdated but informative study of shepherd calls in classical music, see A. Hyatt King, "Mountains, Music, and Musicians," *Musical Quarterly* 31.4 (1945): 395–419. For a wonderful study of parrots, see Paul Carter, *Parrot* (London: Reaktion, 2006); see also the books by Irene Pepperberg discussed later; on birdsong, see, for example, *Birdsong, Speech, and Language: Exploring the Evolution of Mind and Brain*, ed. Johan J. Bolhuis and Martin Everaert (Cambridge, MA: MIT Press, 2013); see also Daniel J. Mennill et. al., "Wild Birds Learn Songs from Experimental Vocal Tutors," *Current Biology* 28.20 (2018): 3273–78; for bird competitions in Suriname, see Emily Hansell Clark, "Iron Gongs and Singing Birds: Paths of Migration and Acoustic Assemblages of Alterity in the Former Dutch Colonial Empire," PhD diss., Columbia University (2020). For more on birds and men in Mozambique, see Claire N. Spottiswoode, Keith S. Begg, and Colleen M. Begg, "Reciprocal Signaling in Honeyguide-Human Mutualism," *Science* 353.6292 (2016): 387–89. The quote about horses is from Vicki Hearne, *Adam's Task: Calling Animals by Name* (New York: Harper Perennial, 1993), 108.

12. I largely skip over the etymological intricacies of the word *communication* in this book, pursuing an approach similar to John Durham Peters (*Speaking into the Air: A History of the Idea of Communication* [Chicago: University of Chicago Press, 1999]), who uses the term *communication theory* "not to refer to any extant practice of inquiry, but in a loose, ahistorical sense for a vision of the human condition as in some fundamental way communicative." (10). That said, Peters does offer a useful etymology of the word, including how it took on its current meaning in English. "'Communication' is a word with a rich history," he observes. "From the Latin *communicare*, meaning to impart, share, or make common, it entered the English language in the fourteenth and fifteenth centuries. The key root is *mun-* (not *uni-*), related to such words as 'munificent,' 'community,' 'meaning,' and *Gemeinschaft*. The Latin *munus* has to do with gifts or duties offered publicly—including gladiatorial shows, tributes, and rites to honor the dead. In Latin, *communicatio* did not signify the general arts of human connection via symbols, nor did it suggest the hope for some kind of mutual recognition. Its sense was not in the least mentalistic: *communicatio* generally involves tangibles. . . . As in Latin, one dominant branch of meaning in 'communication' [in English] has to do with imparting, quite apart from any notion of a dialogic or interactive process. Thus communication can mean partaking, as in being a communicant (partaking of holy communion)" (7). Something closer to how we typically understand communication today—as the sharing of interior ideas or emotions—took root slowly, beginning with Augustine and Aquinas. "But their language was Latin," Peters emphasizes, "and the term *communicatio* itself did not have a privileged role in such discussions. *Communicare* in Latin meant to share or make common and had no special reference to sharing thoughts." It was only in the seventeenth century that the word *communication* in English began to be understood in mentalistic terms (77–78). Peters goes on to show the evolution of the term through the nineteenth century and beyond.

13. Gilbert Ryle, *The Concept of Mind* (New York: Barnes & Noble, 1949).

14. On baboons, see Dorothy L. Cheney and Robert M. Seyfarth, *Baboon Metaphysics: The Evolution of a Social Mind* (Chicago: University of Chicago Press, 2007). On cuttlefish, see Alexandra K. Schnell, Markus Boeckle, Micaela Rivera, Nicola Clayton, and Roger Hanlon, "Cuttlefish Exert Self-Control in a Delay of Gratifica-

tion Task," *Proceedings of the Royal Society B* 288.1946 (2021): 20203161. A vast body of literature exists on cephalopod intelligence, including Peter Godfrey-Smith, *Other Minds: The Octopus, the Sea, and the Deep Origins of Consciousness* (New York: Farrar, Strauss, and Giroux, 2016), as well as texts by Vilém Flusser and Jaron Lanier. On changes in whale song, see Katharine Payne, "The Progressively Changing Songs of Humpback Whales: A Window on the Creative Process of a Wild Animal," in *The Origins of Music*, ed. Nils L. Wallin, Björn Merker, and Steven Brown (Cambridge, MA: MIT Press, 2000); Jenny A. Allen, Ellen C. Garland, Rebecca A. Dunlop, and Michael J. Noad, "Cultural Revolutions Reduce Complexity in the Songs of Humpback Whales," *Proceedings of the Royal Society B* 285.1891 (2018): 20182088.

15. Goodall delivered these remarks as she introduced the "Interspecies Internet Workshop," on 15 July 2019. See http://cba.mit.edu/events/19.07.I2I/.

16. See Irene M. Pepperberg, *Alex & Me: How a Scientist and a Parrot Discovered a Hidden World of Animal Intelligence—and Formed a Deep Bond in the Process* (New York: HarperLuxe, 2008); *The Alex Studies: Cognitive and Communicative Abilities of Grey Parrots* (Cambridge, MA: Harvard University Press, 1999).

17. Monty Roberts, *The Man Who Listens* (New York: Random House, 1997), 101.

18. Piera Filippi et al., "Humans Identify Negative (but Not Positive) Arousal in Silverfox Vocalizations: Implications for the Adaptive Value of Interspecific Eavesdropping," *Current Zoology* 63.4 (2017): 445–56.

19. Patricia B. McConnell and Jeffrey R. Baylis, "Interspecific Communication in Cooperative Herding: Acoustic and Visual Signals from Human Shepherds and Herding Dogs," *Zeitschrift für Tierpsychologie* 67.1–4 (1985): 302–28, 302. See also, Patricia B. McConnell, "Acoustic Structure and Receiver Response in Domestic Dogs, *Canis familiaris*," *Animal Behaviour* 39.5 (1990): 897–904.

20. This work remains highly controversial. See Herbert S. Terrace, *Why Chimpanzees Can't Learn Language and Only Humans Can* (New York: Columbia University Press, 2019).

21. Gabriel delivered a similar talk at a 2013 TED event on the Interspecies Internet. See https://blog.ted.com/the-interspecies-internet-diana-reiss-peter-gabriel-neil-gershenfeld-and-vint-cerf-at-ted2013/.

22. Peter Gabriel, "You know I've worked a lot with musicians from different cultures and part of what you learn is to listen." The event is viewable here: http://cba.mit.edu/events/19.07.I2I/.

23. Jennifer Johung, "Pluripotent Selves and the Performance of Stem Cells," *PUBLIC Journal* 59 (Spring 2019): 171–78. Special issue on "Interspecies Communicated," edited by Meredith Tromble and Patricia Olynyk.

24. Donna Haraway, *The Companion Species Manifesto: Dogs, People, and Significant Otherness* (Chicago: Prickly Paradigm Press, 2003), 35–36.

25. See Peters, *Speaking into the Air*.

26. *Animal Languages: The Secret Conversations of the Living World*, trans. Laura Watkinson (London: John Murray, 2019 [2016]), 12.

27. Peter Wohlleben, *The Hidden Life of Trees: What They Feel, How They Communicate—Discoveries from a Secret World* (Vancouver: Greystone Books, 2015).

28. Eduardo Kohn, *How Forests Think: Toward an Anthropology Beyond the Human* (Berkeley: University of California Press, 2013), chap. 4.

29. Following Manuel DeLanda, I define species as "the historical outcome of

a sorting process (an accumulation of genetic materials under the influence of selection pressures) followed by a process of consolidation (reproductive isolation), which gives a loose accumulation of genes a more or less durable form." Good definitions of the word *species* are hard to come by. For one I've used here, see DeLanda, *A Thousand Years of Nonlinear History* (New York: Zone Books, 1997), 136.

30. Aristotle, *Politics* (New York: Penguin Classics, 1981).

31. Hugh Lofting, *Doctor Dolittle in the Moon*, in *Doctor Dolittle: The Complete Collection*, vol. 4 (New York: Aladdin, 2019 [1928]), 54–55; original emphases.

Chapter 1

1. Looking back on that period, Katharine Boynton Payne made this point explicitly: "There was a burst of realization that the world could change its relation to wildlife. The reaction people had to hearing these sounds made whaling obsolete!" (quoted in David Rothenberg, *Thousand Mile Song: Whale Music in a Sea of Sound* [New York: Basic Books, 2008], 45). On the deployment of whale song in anti-whaling campaigns, and for more on the scientists who cringed at this idea, see D. Graham Burnett, *The Sounding of the Whale: Science and Cetaceans in the Twentieth Century* (Chicago: University of Chicago Press, 2012), 631–36.

2. Roger S. Payne and Scott McVay, "Songs of Humpback Whales," *Science* 173.3997 (1971): 585–97. Tracks 3–5 on the album were recorded by Roger Payne and his former wife, Katherine Boynton (Katy) Payne. The first two tracks were recorded by Frank Watlington, an underwater acoustician working for the US Navy. In the 1960s, Watlington had been stationed on Bermuda and tasked with monitoring the ocean for the sound of Soviet submarines. According to Burnett, "Watlington conveyed almost a decade's worth of tapes to Payne in 1967, when Payne and his wife visited Bermuda. The link to Watlington was made, it appears, by Henry Clay Frick, who had a vacation home in the area" (Burnett, *Sounding of the Whale*, 635, n. 209).

3. Gilbert Grosvenor, quoted in Burnett, *Sounding of the Whale*, 530.

4. Scott McVay, "Can Leviathan Endure So Wide a Chase?" *Natural History* 80.1 (1971): 36–40.

5. I overstate the point. In fact, anti-whaling efforts would continue for the subsequent decades, with important landmark rulings in 1979 and 1982, and commercial whaling continues today in Norway and Japan. Although the situation is less dire than it was in the 1960s, several species of whale remain on the World Wildlife Fund's list of endangered, vulnerable, and threatened animals.

6. Payne and McVay, "Songs of Humpback Whales."

7. Burnett, *Sounding of the Whale*, 629.

8. Roger Payne produced the album. As mentioned, tracks on the album were recorded either by Frank Watlington or by Roger Payne and his wife at the time, Katharine Boynton Payne.

9. I make selective use of these books in what follows: Joe Roman, *Whale* (London: Reaktion Books, 2006); Burnett, *Sounding of the Whale*; Metta Bryld and Nina Lykke, *Cosmodolphins: Feminist Cultural Studies of Technology, Animals, and the Sacred* (London: Zed Books, 2000).

10. Cetacea is actually an infraorder rather than simply an order. Dolphins, in fact, are typically classified as toothed whales, which means that dolphins *are* whales.

11. Gary A. Weir, *An Ocean in Common: American Naval Officers, Scientists, and the Ocean Environment* (College Station: Texas A&M University Press, 2001), especially chap. 16, "Listening, 1946–1961."

12. Bryld and Lykke, *Cosmodolphins*, 19.

13. Burnett, *Sounding of the Whale*; John Durham Peters, *The Marvelous Clouds: Toward a Philosophy of Elemental Media* (Chicago: University of Chicago Press, 2015); and "'Memorable Equinox': John Lilly, Dolphin Vocals, and the Tape Medium," *boundary 2* 47.4 (2020): 1–24.

14. Burnett's *Sounding of the Whale* is an indispensable resource.

15. Burnett, *Sounding of the Whale*, 643–44.

16. This would be a typical maneuver in contemporary media theory, and particularly in media archaeology. For a contextualization and critique of this approach, see Thomas Elsaesser, *Film History as Media Archaeology* (Amsterdam: Amsterdam University Press, 2016).

17. As Burnett observes, Lilly's 1962 article on "Interspecies Communication" in the *McGraw-Hill Yearbook of Science* illustrates "how imprecisely Lilly positioned himself with respect to several significant (and even well-established) disciplines—for instance, the fields of animal behavior, ethology, linguistics, and what Thomas Sebeok and Rulon Wells had recently dubbed 'zoosemiotics' (the study of signaling behavior in and across animal species). From Darwin's *Expression of the Emotions in Man and Animals* to the bee studies of Karl von Frisch and the investigations of Niko Tinbergen and B. F. Skinner there existed not just one but several traditions of research within which Lilly's putatively pioneering investigations fit." The same "imprecision," it seems to me, can be found in all of Lilly's work, including his major texts, such as *Man and Dolphin* and *The Mind of the Dolphin*. Burnett suggests two possible explanations. First, that Lilly had "chosen to willfully disregard the larger disciplinary constellation," which was an inevitable result of "the posturing that must attend any self-conscious 'revolution.'" Second, Lilly "may simply have been genuinely ignorant" of much work in the area. Burnett concludes that "some combination of self-styling strategy and hubristic naïveté seems most probable." Burnett also shows that scientists at the time were well aware of Lilly's ignorance. See Burnett, *Sounding of the Whale*, 588.

18. Burnett points out an overlooked direct connection between Lilly and McVay. McVay worked at Lilly's CRI facility between 1963 and 1965, and it was during that period that he "learned to make and analyze sound spectrograms," which would be so crucial in the famous paper he co-authored with Roger Payne six years later. See Burnett, *Sounding of the Whale*, 635.

19. Ochoa Gautier, "Acoustic Multinaturalism, the Value of Nature, and the Nature of Music in Ecomusicology," in *Econophonia: Music, Value, and Forms of Life*, ed. Gavin Steingo and Jairo Moreno, special issue of *boundary 2* 43.1 (2016): 107–41.

20. John Lilly, *Man and Dolphin: Adventures on a New Scientific Frontier* (New York: Doubleday, 1961), 43.

21. Lilly, *Man and Dolphin*, 46.

22. This second moment is covered brilliantly by John Durham Peters in his article, "'Memorable Equinox'." For another excellent discussion of the basis for Lilly's ideas, see Burnett, *Sounding of the Whale*, 571.

23. Lilly, *Man and Dolphin*, 70. Quotations from this work in the next three paragraphs are cited by page no. parenthetically in the text.

24. Lilly, *Man and Dolphin*, 75. This was the day, says Peters, that Lilly "hit the phonocentric jackpot." See Peters, "'Memorable Equinox.'"

25. Lilly, *Man and Dolphin*, 75. Lilly filled in some of the gaps in his 1988 book, *The Scientist: A Metaphysical Autobiography* (Berkeley, CA: Ronin Publishing, 1988), a bizarre autobiography written in the third person.

26. "The absolute size of a mammalian brain determines its computing capability and the size of its storage (memory); the larger the computer, the greater its power." See John Lilly, *The Mind of the Dolphin: A Nonhuman Intelligence* (New York: Avon, 1967), 33; cf. Lilly, "Critical Brain Size and Language," *Perspectives in Biology and Medicine* 6 (1963): 246–55. Later in his career, Lilly systematized his human-computer analogy with the concept of the "human biocomputer." See Lilly, *Programming and Metaprogramming in The Human Biocomputer* (New York: Julian Press, 1967). Contemporary neuroscientists balk at such facile comparisons. As Peters points out, "Whales have the biggest brains of any animal on earth in terms of sheer size, with sperm whale brains weighing in at more than eight kilograms, but absolute brain size is a poor indicator of intelligence. A better measure is the encephalization quotient (EQ), a statistical measure comparing total brain size to expected total body volume, and dolphins rank second of all creatures on earth in EQ, above the great apes but below human beings." See Peters, *Marvelous Clouds*, 59.

27. Lilly, *Mind of the Dolphin*, 83, 108–9.

28. Lilly, *Mind of the Dolphin*, 198.

29. Lilly, *Mind of the Dolphin*, 86.

30. See Bruce Clarke, "John Lilly, The Mind of the Dolphin, and Communication Out of Bounds," *communication+1* 3 (2014); http://scholarworks.umass.edu/cpo/vol3/iss1/8/.

31. Lilly, *Man and Dolphin*, 22–23.

32. Lilly, *Man and Dolphin*, 180.

33. I thank Manuel DeLanda for helping me with this description.

34. The auditory apparatus of the dolphin, in contrast, is distributed across the animal's head. Sounds are channeled through the "melon" (a large fatty mass in the forehead that is often compared to an "acoustic lens") and specialized tissue in the jaw before being transmitted to the inner ear.

35. Lilly, *Man and Dolphin*, 198.

36. Richard Eindhorn, "Dolphins Challenge the Designer," *Electronic Design* 15.25 (1967): 49–64, 52.

37. "Agreement among listeners ruled out 'imagination,'" he hastily interjected, although his defensiveness on this point suggests that he may not have been entirely convinced. See Lilly, *Mind of the Dolphin*, 134.

38. Lilly, *Mind of the Dolphin*, 134.

39. The American Psychological Association's *Dictionary of Psychology* provides a description of "difference tone" under the entry for "combination tone," which it defines as "a tone generated in the ear that is produced when two primary tones (i.e., two tones differing in frequency) are presented simultaneously. For example, under certain conditions a 1000 Hz tone and a 1200 Hz tone will produce audible combination tones whose frequencies are 200 Hz (a difference tone) and 800 Hz (a cubic

difference tone). Most combination tones are produced by nonlinear distortion that occurs within the cochlea." Difference tones are subjective in the sense that they are produced in the human ear, unlike most frequencies we associate with sound (including fundamental frequencies and overtones), which are physical vibrations in a medium such as air. See the entry at https://dictionary.apa.org/combination-tone.

40. Lilly, *Mind of the Dolphin*, 136.

41. He would later comment on this epiphany that human/dolphin communication was first and foremost a problem of bodies: "He [i.e., Lilly] realized, confined in the human body, his difference from the dolphins confined in their bodies" (Lilly, *The Scientist*, 132).

42. Lilly, *Mind of the Dolphin*, 82.

43. Lilly, *Mind of the Dolphin*, 83.

44. *The Girl Who Talked with Dolphins* [film], directed by Christopher Riley (London: BBC, 2014).

45. Human Dolphin Foundation brochure, John C. Lilly Papers, Stanford University (box 38, folder "The Human Biocomputer"). "Through JANUS, both man and dolphin speak by means of meaning-assigned whistles. The computer generates the human side of conversation from keyboard input, and the dolphins learn to understand and send the whistles. The computer then deciphers the dolphin-sent JANUS language whistles and displays messages on a CRT [i.e., cathode ray tube]. JANUS is one of three communication projects ongoing at HDF [i.e., Human Dolphin Foundation]." See Leslie Baer-Frohoff, "The Interspecies Connection," *HardCopy* 13.11 (1984): 44. The Human Dolphin Foundation was founded in 1976 by John C. Lilly, Antionetta Lilly, Burgess Meredith, and Victor di Suvero.

46. Human Dolphin Foundation brochure, John C. Lilly Papers, Stanford University (box 38, folder "The Human Biocomputer").

47. Lilly, *Mind of the Dolphin*, 70, 88, 90.

48. "Dolphins Challenge the Designer," 60. Lilly also used an Apple II system donated to him directly by none other than Steve Jobs. Letter to Steve Jobs from John Kert (Director of Research, Human Dolphin Foundation), John C. Lilly Papers, Stanford University (box 1, folder: "Apple Computers").

49. See Peters, "'Memorable Equinox'" for an excellent account of Lilly' deployment of tape.

50. Lilly, *Man and Dolphin*, 80.

51. In media theory, this process is called "time axis manipulation." People often assume that pitch and speed were *inherently* linked in the analog era and that the separation of those parameters became possible only through digital technologies. This assumption is not entirely correct, however, as forthcoming work by Jonathan Sterne and Mara Mills shows clearly.

52. Roger Payne, *Among Whales* (New York: Delta, 1995), 21.

53. David Rothenberg (words) and Michael Deal (images), "Whale Song Explained," online at *Medium*, 10 October 2014; https://medium.com/@dealville/whales-synchronize-their-songs-across-oceans-and-theres-sheet-music-to-prove-it-b1667f603844.

54. Rothenberg, "Whale Song Explained."

55. Lilly, "A Feeling of Weirdness," in *Mind in the Waters: A Book to Celebrate the Consciousness of Whales and Dolphins*, ed. Joan McIntyre (New York: Scribner, 1974), 74.

56. Lilly, *Man and Dolphin*, 28.

57. Lilly, *Mind of the Dolphin*, 155.

58. This narrative has been contested; see Burnett, *Sounding of the Whale*, 612, n. 160.

59. See or hear the interview with Howe in "Home Is Where Your Dolphin Is," *Radiolab* (21 August 2014); https://www.radiolab.org/episodes/home-where-your-dolphin.

60. Howe's account is printed in Lilly's *Mind of the Dolphin*, 219–20.

61. Lilly, *Mind of the Dolphin*, 198.

62. Lilly, *Man and Dolphin*, 122.

63. The word "commensurate" is from Burnett's *Sounding of the Whale*, 580.

64. Lilly, *Mind of the Dolphin*, 155.

65. Lilly, *Mind of the Dolphin*, 158.

66. Lilly, *Man and Dolphin*, 110.

67. See Peters, *Marvelous Clouds*, chap. 2, for an excellent analysis of this aspect of Lilly's work.

68. Like many mammals, dolphins do use tools. See, for example, Sonja Wild, et al., "Multi-Network-Based Diffusion Analysis Reveals Vertical Cultural Transmission of Sponge Tool Use within Dolphin Matrilines," *Biological Letters* 15.7 (2019); http://dx.doi.org/10.1098/rsbl.2019.0227.

69. Lilly, *Mind of the Dolphin*, 55.

70. Lilly, *Mind of the Dolphin*, 121.

71. See Bryld and Lykke *Cosmodolphins*, 63, for more on this point.

72. Douglas Adams, *The Hitchhiker's Guide to the Galaxy* (New York: Harmony Books, 1980), 156.

73. In the Soviet Union, too, awareness of cetacean intelligence occasionally led to ideological crises—if genuine equality can exist in the absence of tools, perhaps the Marxist emphasis on industrial production is wide off the mark. As one Soviet writer in the early 1970s speculated about dolphin socialization, "And did not this non-technological society seriously question the Marxist-Leninist belief in the inevitability of the industrial revolution as a global phenomenon?" (Konstantinov, quoted in Bryld and Lykke, *Cosmodolphins*, 169).

74. Sidney Cohen, quoted in John Lilly, *The Center of the Cyclone: An Autobiography of Inner Space* (New York: The Julian Press, 1972), 141.

75. Lilly, *The Deep Self: Profound Relaxation and the Tank Isolation Technique* (New York: Simon and Schuster, 1977), 71.

76. Lilly, *The Deep Self*, 77, 86.

77. Lilly, *The Deep Self*, 102.

78. Lilly, *The Deep Self*, 106.

79. Lilly, *The Scientist*, 132.

80. I thank Roger Mathew Grant for finding this document (which we refer to jokingly as Lilly's "LSD Diary") among the Lilly Papers. See John C. Lilly Papers, Stanford University (box 52, folder 8). Lilly mentions his visit, as well as the journal, in *Center of the Cyclone* (147).

81. In *The Center of the Cyclone*, Lilly writes: "*Bolero*, according to Idries Shah as well as Oscar Ichazo, is a piece of Sufi music composed by the Chusti group of Sufis to induce special states of consciousness" (125).

82. Lilly, *Center of the Cyclone*, 10

83. Lilly, *Mind of the Dolphin*, 112. Since Mahler never wrote a piece with this title, it is unclear what Lilly has in mind here. He may have been thinking of Mahler's "Titan" Symphony.

84. Once again, this was a "hypothesis" based purely on faith (Lilly, *Mind of the Dolphin*, 112).

85. Antonia Majaca and Luciana Parisi, "The Incomputable and Instrumental Possibility," *E-Flux* 77 (November 2016); https://www.e-flux.com/journal/77/76322/the-incomputable-and-instrumental-possibility.

86. Ochoa Gautier, "Acoustic Multinaturalism," 127. The phrase "the immaterialization of capital" is from George Yúdice, *The Expediency of Culture: Uses of Culture in the Global Era* (Durham, NC: Duke University Press, 2003), 9.

87. Roger Mathew Grant, "Music Lessons on Affect and Its Objects," *representations* 144.1 (2018): 34–60.

88. Forkel, quoted in Grant, "Music Lessons," 43.

89. Grant, "Music Lessons," 49.

90. Joan McIntyre, "Mind Play," in *Mind in the Waters: A Book to Celebrate the Consciousness of Whales and Dolphins*, ed. Joan McIntyre (New York: Scribner, 1974), 94.

91. Bryld and Lykke, *Cosmodolphins*, 18.

92. Helmreich, *Sounding the Limits of Life: Essays in the Anthropology of Biology and Beyond* (Princeton, NJ: Princeton University Press, 2016), 145, my emphasis.

93. Sigmund Freud, *Das Unbehagen in der Kultur* (Vienna: Internationaler Psychoanalytischer Verlag, 1930).

94. This attitude is apparent already in the early 1960s, although not earlier than that. See W. E. Schevill, "Whale Music," *Oceanus* 9.2 (1962): 3–13. Burnett writes, "As for the notion of these creatures possessing 'intelligence' or 'beauty,' one must comb closely through marginal materials to catch even a hint of such preoccupations in the later 1950s" (*Sounding of the Whale*, 522).

95. Rothenberg, "Whale Song Explained."

96. Roger Payne, quoted in Tim Lewis, "'It Always Hits Me Hard': How A Haunting Album Helped Save the Whales," *Guardian*, 6 December 2020; https://www.theguardian.com/environment/2020/dec/06/the-songs-that-saved-the-whales.

97. A note on my use of the word *discovered*: It is imperative to reject the antirealist position, all too common in recent academic writing, expressed, for example, in a recent paper in the journal *Environment and Planning*: "Whale music was not discovered, as many of its devotees proposed it was, but invented, through a combination of animal sounds, recording techniques, consumer trends, and pervasive beliefs about nature." (See Max Ritts, "Environmentalists Abide: Listening to Whale Music, 1965–1985," *Environment and Planning D: Society and Space* 35.6: 1109.) It is time to break with this kind of strong constructivism and to recognize that the repeated structure of whale vocalization cannot be reduced to human invention. To be sure, the mere fact of this structure says nothing about the gymnastics required to graft whale "song" onto an ethical grammar of suffering, nor does it mean a priori that the structured vocalizations are in fact songs in any meaningful sense. But the work of Payne, McVay, and many others does tell us many things about whales that we previously did not know. The highly formalized structure of whale vocalization implies cognitive abilities that had previously not been recognized by humans. Perhaps one might put it like this: until the 1960s, few people (in the West, at least) were paying attention

to the sound of whales. The few who did directly encounter whale phonation came upon it accidently and through an entirely different agenda: surveying oceans for the purposes of marine warfare. When people started paying serious attention to whale vocalization, they used the concepts available to them to understand what they were hearing. This was less an invention than a way of making sense of the phenomenon of patterned whale vocalization, which previously had been practically ignored. I thank Anthony Reed for helping me with this formulation.

98. Payne and McVay, "Songs of Humpback Whales," 590.

99. Margret Grebowicz, *Whale Song* (New York: Bloomsbury, 2017), 2.

100. W. B. Broughton, quoted in Payne and McVay, "Songs of Humpback Whales," 590.

101. It was not until the early twenty-first century that a team of researchers led by Rebecca Dunlop undertook a systematic study of what they call "non-song acoustic communication," or "social sounds." See, for example, Rebecca A. Dunlop, Douglas H. Cato, and Michael J. Noad, "Non-Song Acoustic Communication in Migrating Humpback Whales (*Megaptera novaeangiae*)," *Marine Mammal Science* 24.3 (2008): 613–29.

102. In the short text (actually a "glossarial index") to which Payne and McVay refer, Broughton begins: "W. H. Thorpe, in his recent monograph, *Bird Song* (1961), grants that the title *Bird Vocalizations* would be both more accurate and more objective, because of the anthropocentric artistic connotations of the word *song*. Wisely, however, he avoids the *reductio ad absurdum* of the ultimate unattainable in objectivity, and sticks to the familiar word despite its disadvantages" (W. B. Broughton, "Glossarial Index," in *Acoustic Behavior of Animals*, ed. R.-G. Busnel [Amsterdam: Elsevier, 1963], 882). Objectivity is one thing, but meaningful terms that illuminate rather than mislead are another. There is no *reductio ad absurdum* in opting for better terms.

103. Eduardo Mercado III, "Acoustic Signaling by Singing Humpback Whales (*Megaptera novaeangliae*): What Role Does Reverberation Play?" *PLoS ONE*, 1 December 2016; https://journals.plos.org/plosone/article?id=10.1371/journal.pone.0167277.

104. My presentation here is a simplified version of what Peters presents; see his *Marvelous Clouds*, esp. 389–91.

105. See Peters, *Marvelous Clouds*, 94.

106. Peters, *Marvelous Clouds*, 95.

107. Peters, *Speaking into the Air*, 65.

108. Roger Payne, quoted in Lewis, "'It Always Hits Me Hard.'"

Chapter 2

1. Zakes Mda, *The Whale Caller* (London: Viking, 2005). Portions of this chapter were previously published as "Whale Calling," *Ethnomusicology* 65.2 (2021): 350–73. Used with permission from the Society for Ethnomusicology.

2. Percival Kirby, *The Musical Instruments of the Native Races of South Africa* (Oxford: Oxford University Press, 1934), 83.

3. Mda, *Whale Caller*, 57.

4. Mda, *Whale Caller*, 57.

5. Mda, *Whale Caller*, 59.

6. Mda, *Whale Caller*, 57. Mda writes that the sound "wavers between the military-inflected arpeggiations of a bugle call and the fast-paced scatting of a demented jazz singer" (59).

7. Mda, *Whale Caller*, 59.

8. Mda, *Whale Caller*, 11.

9. To be more precise, and taking a cue from animal studies scholars, the distinction would really be one between humans and *non-human* animals, since humans are animals after all.

10. Mda, *Whale Caller*, 35.

11. Angelique Serrao, "Conveyor of Magic: Interview with Zakes Mda," *Saturday Star* [Johannesburg], 24 June 2006, 5; as quoted in Harry Sewall, "Border Crossings: Mapping the Human and the Nonhuman in Zakes Mda's *The Whale Caller*," *Scrutiny2* 12.1 (2007): 129–38, 130.

12. While both Mda and Salukazana were born in the Eastern Cape in Xhosa-speaking communities, the former grew up in a middle-class family and moved to Lesotho at an early age. For more on Mda, see his *Sometimes There Is a Void: Memoirs of an Outsider* (New York: Farrar, Straus and Giroux, 2012).

13. The camp is called Habonim Dror.

14. Eric Davalala was born in 1989 in the small town of Willowvale in the Eastern Cape, nearly seven hundred miles from Hermanus. He moved to Hermanus in 2009.

15. The quotations from Eric Davalala in this chapter are from an interview I recorded and transcribed with his permission.

16. On apartheid townships as "dormitory towns," see Jane Starfield and Michael Gardiner, "Citizenship and Modernity in Soweto," in *Senses of Culture: South African Cultural Studies*, ed. Sarah Nuttall and Cheryl-Ann Michael (Cape Town: Oxford University Press, 2001), 61–84, 69.

17. Wilson Salukazana permitted me to record our conversation and publish it because, as he told me, he wants people around the world to know about Hermanus and its whales. My interview with Salukazana was conducted in English, with smatterings of Xhosa throughout. Thanks to Luvo Ntliziywana for checking my Xhosa translations.

18. This "code" was developed by the Hermanus tourist office. Although more of a gimmick than anything else, the idea is that the whale crier blows the horn in specific long/short configurations to notify tourists of a whale's location. A long blow followed by two staccato notes means that whales have been sighted at New Harbour, for example, while four staccato notes in quick succession indicates Roman Rock. The system is detailed on a sandwich board the whale crier wears during working hours.

19. Born Again (also known as Reborn) Christian refers to one of the many branches of evangelical Christianity that have long existed in South Africa but which have become much more prominent in recent years. By specifying that Wilson is a Born Again Christian, Philemon implied a particular form of Christian worship that often sees itself as incommensurable with indigenous African practices.

20. Nicholas A. Smyrnios and Richard S. Irwin, "Current Concepts in the Pathophysiology and Management of Near-Drowning," *Journal of Intensive Care Medicine* 6.1 (1991): 26–35, 28.

21. M. J. Tipton, "Editorial Review: The Initial Responses to Cold-Water Immersion in Man," *Clinical Science* 77 (1989): 581–88, 581.

22. Tipton, "Editorial Review," 582.

23. Dorothy Winifred Steele, "Interpreting Redness: A Literary Biography of Zakes Mda" (MA thesis, University of South Africa, 2007), 265–66. See also Mda, *Sometimes There Is a Void*, 509.

24. O. B. Lawuyi, "Water, Healing, Gender and Space in African Cosmology," *South African Journal of Ethnology* 21.4 (1998): 185–90.

25. Manton Hirst, "A River of Metaphors: Interpreting the Xhosa Diviner's Myth," *African Studies* 56.2 (1997): 217–50. Hirst notes that *ukuthwetyulwa* is related to another part of the process known as *ukuthwasa*, which means "to appear gradually."

26. Joan Broster, *Amagqirha: Religion, Magic and Medicine in Transkei* (Goodwood: Via Africa, 1981), 47.

27. Hirst, "River of Metaphors," 234. Felicity Wood writes that "in terms of the beliefs associated with the initiation of a Xhosa *igquirhas* (diviner), part of the process is said to involve a descent into a river where an ancestor figure, part human and part fish, instructs the initiate." See Wood, "The Snake in the Sky: Tornadoes in Clay and Local Narrative in the Hogsback-Alice Area," *Critical Arts* 14.2 (July 2000): 79–95, 87.

28. Katesa Schlosser, *Zauberei im Zululand: Manuskripte des Blitz-Zauberers Laduma Madela* (Kiel: Schmidt & Klauning, 1972), 255. Translations from German by the author.

29. Mda, *Sometimes There Is a Void*, 4.

30. Mda, *Sometimes There Is a Void*, 4.

31. Wood, "Snake in the Sky," 87.

32. For information on all these aspects of Mda's life, see his memoir, *Sometimes There Is a Void*.

33. hooks, *All About Love: New Visions* (New York: Harper, 2000).

34. The entries were published in Lilly's *The Mind of the Dolphin: A Nonhuman Intelligence* (New York: Avon, 1967).

35. Lilly, *Mind of the Dolphin*, 83.

36. Lilly, *Mind of the Dolphin*, 198.

37. Entries in Lilly, *Mind of the Dolphin*, 231, 233–34, 241, 256–57, 259.

38. *The Girl Who Talked with Dolphins* [film], directed by Christopher Riley (London: BBC, 2014).

39. Christopher Riley, "The Dolphin Who Loves Me: The NASA-Funded Project That Went Wrong," *The Guardian*, 8 June 2014; https://www.theguardian.com/environment/2014/jun/08/the-dolphin-who-loved-me.

40. Lilly, *Mind of the Dolphin*, 155.

41. *Mind of the Dolphin*, 20, my emphasis. In 1962, the sexologist R. E. L. Masters argued that human sex with an animal of the opposite sex is less "perverted" than homosexual human sex because in the former case bestialists "humanize" their "opposite-sex animal love object" while the love object of the homosexual is an abnormal one (and by implication homosexuals "animalize" humans). See Joanna Bourke (*Loving Animals: On Bestiality, Zoophilia and Posthuman Love* [London: Reaktion, 2020], 88) for an excellent critique of the medicalization and pathologization of bestiality in relation to other forms of "deviant" sexuality.

42. Lilly, *Mind of the Dolphin*, 155.

43. Lilly, *Mind of the Dolphin*, 155.

44. Lilly, *Mind of the Dolphin*, 155.

45. Amia Srinivasan, "What Does Fluffy Think? (Review of *Loving Animals: On Bestiality, Zoophilia and Posthuman Love*, by Joanna Bourke)," *London Review of Books* 43.19 (2021); https://www.lrb.co.uk/the-paper/v43/n19/amia-srinivasan/what-does-fluffy-think.

46. The article is Gieri Bolliger and Antoine Goetschel, "Sexual Relations with Animals (Zoophilia): An Unrecognised Problem in Animal Welfare Legislation," in *Bestiality and Zoophilia: Sexual Relations with Animals*, ed. Andrew Beetz and Anthony Podberscek (West Lafayette, IN: Purdue University Press, 2005), 23–45.

47. See Srinivasan, "What Does Fluffy Think?" She quotes Bolliger and Goetschel, "Sexual Relations with Animals," 39.

48. Lilly, *Man and Dolphin: Adventures on a New Scientific Frontier* (New York: Doubleday, 1961), 138–39.

49. Lilly, *Man and Dolphin*, 147.

50. Peter Singer, "Heavy Petting," *Prospect*, April 19, 2001; https://www.prospectmagazine.co.uk/opinions/56258/heavy-petting.

51. Laurie Bergman, "Canine Mounting: An Overview," *Clinician's Brief*, January 2012, 61–63.

52. Bateson, "A Theory of Play and Fantasy," in *Steps to an Ecology of Mind* (Chicago: University of Chicago Press, 2000), 180; cited in Kohn, *How Forests Think: Toward an Anthropology Beyond the Human* (Berkeley: University of California Press, 2013), 147.

53. Pelin Sahlén, Liu Yanhu, Jinrui Xu, Eniko Kubinyi, Guo-Dong Wang, and Peter Savolainen, "Variants That Differentiate Wolf and Dog Populations Are Enriched in Regulatory Elements," *Genome Biology and Evolution* 13.4 (2021); https://doi.org/10.1093/gbe/evab076.

54. Wolves do, however, exhibit these traits in the presence of other wolves. See Friederike Range and Zsófia Virányi, "Tracking the Evolutionary Origins of Dog-Human Cooperation: The 'Canine Cooperation Hypothesis,'" *Frontiers in Psychology* 5 (January 15, 2015); https://doi.org/10.3389/fpsyg.2014.01582.

55. Although debate rages about the date(s) and location(s) of dog domestication, the consensus view holds that domestication "began when wolves began following human hunting parties to scavenge for leftovers." Richard C. Francis provides an excellent and very readable account. See his *Domesticated: Evolution in a Man-Made World* (New York: W. W. Norton, 2015), 27. Chapters 1 and 2 are particularly relevant.

56. Peter Saetre et al. "From Wild Wolf to Domestic Dog: Gene Expression Changes in the Brain," *Molecular Brain Research* 2004 (126): 198–206; 10.1016/j.molbrainres.2004.05.003.

57. On gene expression in dogs and wolves, see Saetre et al. On noncoding aspects of genomes, see Sahlén et al.

58. Francis writes, "The amount of change in phenotype (which includes behavioral, physiological, and morphological traits) does not correlate closely with changes in genome (genetic makeup). For example, large physical changes in domesticated animals often involve remarkably few genetic alterations. The genetic distance between dogs and wolves is tiny compared to their physical distance." See *Domesticated*, 5.

59. Negar Didehban, Mahdi Pourmahdi Borujeni, Reza Avizeh, and Bahman

Mosallanejad, "Problematic Behaviors in Companion Dogs: A Survey of Their Prevalence and Associated Factors," *Journal of Veterinary Behavior* 39 (2020): 6–13, 11.

60. Gilles Deleuze and Félix Guattari, *A Thousand Plateaus: Capitalism and Schizophrenia*, trans. Brian Massumi (Minneapolis: University of Minnesota Press, 1987), 240; emphasis in original.

61. See the discussion of "animal" in the film, *Gilles Deleuze from A to Z*, dir. Pierre-André Boutang (New York: Semiotext(e), 2011).

62. Donna Haraway, *The Companion Species Manifesto: Dogs, People, and Significant Otherness* (Chicago: Prickly Paradigm Press, 2003), 5.

63. Haraway, *Companion Species Manifesto*, 33.

64. Haraway, *Companion Species Manifesto*, 33.

65. Weisser, as quoted in Haraway, *Companion Species Manifesto*, 37–38. Haraway quotes from the Great Pyrenees Discussion List from 2001 and 2002.

66. Haraway, *Companion Species Manifesto*, 95–96.

67. "Kin making is making persons, not necessarily as individuals or as humans." See Donna Haraway, *Staying with the Trouble: Making Kin in the Chthulucene* (Durham, NC: Duke University Press, 2016), 103.

68. Deleuze and Guattari, *A Thousand Plateaus*, 233.

69. Haraway, *Companion Species Manifesto*, 3.

70. Ted Chiang, "The Lifecycle of Software Objects," in *Exhalation* (New York: Alfred A. Knopf, 2019). The author's note is only available in the audiobook version. I thank Steven Chung for introducing me to Chiang's work.

71. hooks, *All About Love: New Visions* (New York: Harper, 2000).

72. Irene Pepperberg, *Alex & Me: How a Scientist and a Parrot Discovered a Hidden World of Animal Intelligence—and Formed a Deep Bond in the Process* (New York: HarperLuxe, 2008), 8. Hereafter cited in the text by page no. in parentheses (except for n. 75).

73. All quotations in this paragraph are from Pepperberg, *Alex & Me*, 220.

74. The description of *taiko* (a style of Japanese and Asian-American drumming) as "atonal" is perplexing, for example, and the characterization of African music as having "almost all strong beats" is equally troublesome (see p. 504). Atonality usually refers to modernist music in the late nineteenth and early twentieth centuries written by composers deliberately breaking down key-centered diatonic structures—hardly a description of *taiko*. It's possible the authors are trying to say that *taiko* emphasizes non-pitched percussion instruments, and is therefore "atonal" in the sense of having nothing to do with tonality. But that's just a speculation on my part. The comment about Akan music is contradicted by most of what we know about metrical structure in music. See M. E. Mingle et al., "Chimpanzees Prefer African and Indian Music Over Silence," *Journal of Experimental Psychology: Animal Learning and Cognition* 40.4 (2014): 502–50.

75. Gavin Maxwell, *Ring of Bright Water* (London: Longmans, Green, 1960).

76. Werner Herzog at Princeton University, April 6, 2022.

77. None of this needs to be taken obviously or literally: ancestors might simply mean a storehouse of historical knowledge. This is not a topic I can adequately address here, but many understand ancestral knowledge to be a form of connection to the past that allows for acute awareness in the present. In this sense, our ancestors "literally" are with us. As Gabriel Rockhill says, our bodies decay, but we never really

die. See Rockhill, "Why We Never Die," *New York Times*, 29 August 2016; https://www.nytimes.com/2016/08/29/opinion/why-we-never-die.html.

78. M. Sakai, Y. Kita, K. Kogi, et al., "A Wild Indo-Pacific Bottlenose Dolphin Adopts a Socially and Genetically Distant Neonate," *Scientific Reports* 6 (2016); https://www.nature.com/articles/srep23902; cited in Alexis Pauline Gumbs, *Undrowned: Black Feminist Lessons from Marine Mammals* (Chico, CA: AK Press, 2020), 161–62, n. 50.

79. Gumbs, *Undrowned*, 162.

80. Margret Grebowicz, *Whale Song* (New York: Bloomsbury, 2017), 40.

81. Gumbs, *Undrowned*, 8.

82. Gumbs, *Undrowned*, 71.

83. Gumbs, *Undrowned*, 8, 20, 22, 22, 30, 22.

84. Grebowicz, *Whale Song*, 111.

85. Roland Barthes, *A Lover's Discourse: Fragments*, trans. Richard Howard (New York: Hill and Wang, 1978), 148.

86. Part of that trespassing is deflating the fake grandiosity of White thought—the kind of thinking that would precisely place Nietzsche, Lacan, Hölderlin, Proust, and the human, on the one side, and everything else on the other.

87. Barthes, *A Lover's Discourse*, 149.

88. John Durham Peters, *Speaking into the Air: A History of the Idea of Communication* (Chicago: University of Chicago Press, 1999), 180.

Prefatory Remark to the Second Part

1. See, for example, Philippe Descola, *Beyond Nature and Culture*, trans. Janet Lloyd (Chicago: University of Chicago Press, 2013).

2. Michael Schetsche, René Gründer, Gerhard Mayer, and Ina Schmied-Knittel, "Der maximal Fremde: Überlegungen ze einer transhumanen Handlungstheorie," *Berliner Journal für Soziologie* 3 (2009): 469–91.

3. See, for example, Michael J. Crowe, *The Extraterrestrial Life Debate, 1750–1900: The Idea of a Plurality of Worlds from Kant to Lowell* (Cambridge: Cambridge University Press, 1986).

4. See https://breakthroughinitiatives.org/initiative/1.

Chapter 3

1. Frank Drake, "Prospects in the Search for Extraterrestrial Civilizations," paper presented at the American Astronautical Society, 12th Annual Meeting, Anaheim, California, 23–25 May 1966, 66–78.

2. John C. Lilly Papers, Stanford University, box 28.

3. Metta Bryld and Nina Lykke, *Cosmodolphins: Feminist Cultural Studies of Technology, Animals, and the Sacred* (London: Zed Books, 2000), chap. 7 ("Voices from Inner and Outer Space").

4. On the complexities of the translation and its interpretation, see Michael J. Crowe, *The Extraterrestrial Life Debate, 1750–1900: The Idea of a Plurality of Worlds from Kant to Lowell* (Cambridge: Cambridge University Press, 1986), 14–16.

5. Loren Eiseley, "The Long Loneliness," in *The Star Thrower* (New York: Harcourt Books, 1978), 37–44, 37.

6. See Steven J. Dick, *Astrobiology, Discovery, and Societal Impact* (Cambridge: Cambridge University Press, 2018).

7. Carl Sagan, *Cosmos* (New York: Random House, 1980), 287.

8. Epicurus, "Letter to Herodotus," trans. C. Bailey, in *The Stoic and Epicurean Philosophers*, ed. Whitney J. Oates (New York, 1957), 3–15, 5, 12.

9. These are prolific authors. A small but important selection of their work includes the following: Crowe, *Extraterrestrial Life Debate, 1750–1900*; Steven J. Dick, *Plurality of Worlds: The Extraterrestrial Life Debate from Democritus to Kant* (Cambridge: Cambridge University Press, 1982); Guthke, *Der Mythos der Neuzeit: Das Thema der Mehrheit der Welten in der Literatur- und Geistegeschichte von der kopernikanischen Wende bis zur Science Fiction* (Bern: Francke, 1983). Also useful is Michael J. Crowe, ed., *The Extraterrestrial Life Debate, Antiquity to 1915: A Source Book* (Notre Dame, IN: University of Notre Dame Press, 2008).

10. "Many intellectuals in the first half-dozen decades of the seventeenth century seem to have been so confused, so undecided, or so cautious about the question of a plurality of inhabited worlds that scholars have had difficulty specifying the actual views they held" (Crowe, *Extraterrestrial Life Debate, Antiquity to 1915*, 14).

11. Most important was Descartes's model of considering each star as a sun surrounded by planets. See M. de Fontenelle (Bernard Le Bovier), *Conversations on the Plurality of Worlds*, trans. Elizabeth Gunning (London: J. Cundee, 1803); orig. pub. in French as *Entretiens sur la pluralité des mondes* (1686).

12. Fontenelle, *Conversations*, 85.

13. Fontenelle, *Conversations*, 51.

14. "The objection then turns on the existence of men in the moon, but it is the objectors themselves who talk of men [*hommes*] as its inhabitants; I have asserted no such thing: I say there are inhabitants [*habitants*], and I likewise say they may not at all resemble us" (Fontenelle, *Conversations*, xv).

15. Leibniz, quoted in Crowe, *Extraterrestrial Life Debate, 1750–1900*, 28–30.

16. David Grinspoon, *Lonely Planets: The Natural History of Alien Life* (New York: Ecco, 2003), 15.

17. The Copernican Revolution, after all, happened almost simultaneously with the "discovery" of the Americas. In the fifteenth century, Christian scholars conceived the habitable world as a single insular mass (something like an island), surrounded by unknowable and uninhabitable waters. This insular mass (which included bits of Europe, Asia, Africa, and the Atlantic) had the holy city of Jerusalem at its center "and the night of the Ocean all around." After Columbus's voyage, however, the notion of the *Orbis terrarium* was called into question: the Ocean no longer demarcated the limits of the world, but was reduced to little more than a border, such as a river or mountain range that separates provinces. See Bernard McGrane, *Beyond Anthropology: Society and the Other* (New York: Columbia University Press, 1989), 32.

18. Ana María Ochoa Gautier, *Aurality: Listening and Knowledge in Nineteenth-Century Colombia* (Durham, NC: Duke University Press, 2014), 124.

19. Locke, *An Essay Concerning Human Understanding* (Dublin: R. Milliken, 1825 [1689]), bk. 4, 425.

20. Locke, *An Essay Concerning Human Understanding*, 425–26.

21. For an excellent reading, see Peter Szendy, *Kant in the Land of the Extraterrestrials: Cosmopolitical Philosofictions*, trans. Will Bishop (New York: Fordham University Press, 2013).

22. Kant, *Anthropology from a Pragmatic Point of View*, trans. and ed. Robert B. Louden (Cambridge: Cambridge University Press, 2006 [1798]), 225.

23. This problem works against the grain of another gesture in Kant's *Anthropology* that would define the human as above (i.e., ontologically higher than) the animal. The comparison—or rather, non-comparison—with the extraterrestrial, is "*horizontal* rather than vertical, in the sense that it seems a priori to imply none of the hierarchies that traditionally situate the human as a mortal and a rational animal." Szendy, *Kant in the Land of the Extraterrestrials*, 142; emphasis in original.

24. Note the etymological and conceptual connection between "solution" as a problem to be solved, on the one hand, and "solution" as liquid, on the other. To say that something is insoluble (unsolvable, *unauflöslich*) means at once that it cannot be answered (solved) and that it cannot become a liquid (dissolve). To solve the problem of humanity, in other words, means to make it soluble (*auflöslich*)—so that it dissolves and flows away like water.

25. For a useful discussion about whether Carl Friedrich Gauss tried something earlier, see Crowe, *Extraterrestrial Life Debate, 1750–1900*, 205.

26. Charles Cros, quoted in Crowe, *Extraterrestrial Life Debate, 1750–1900*, 359.

27. Brett Brehm, "Paleophonic Futures: Charles Cros' Audiovisual Worlds," *Nineteenth-Century French Studies* 45.3–4 (2017): 179.

28. Jonathan Sterne, *The Audible Past: Cultural Origins of Sound Reproduction* (Durham, NC: Duke University Press, 2003), 80, 35. Thomas Y. Levin similarly observes that Cros came up with "the principle for a machine that accomplished both inscription and subsequent reproduction." He attributes Edison's success over Cros to the latter's poverty: "The only problem was that the poor poet Cros—a friend of Verlaine's, and favorite son of the Surrealists—was just that: poor. He thus did not have the money to actually build the device." See Levin, "Before the Beep: A Short History of Voice Mail," in *Voice: Vocal Aesthetics in Digital Art and Media*, ed. Theo van Leeuwen, Ross Gibson, and Norie Neumark (Cambridge, MA: MIT Press, 2010 [1995]): 17–32, 25.

29. "Marconi Sure Mars Flashes Messages," *New York Times*, 2 September 1921, 1.

30. Crowe, *Extraterrestrial Life Debate, 1750–1900*, 458.

31. Leonor Sierra, "Are We Alone in the Universe: Revisiting the Drake Equation," *Exoplanet Exploration*, 19 May 2016; https://exoplanets.nasa.gov/news/1350/are-we-alone-in-the-universe-revisiting-the-drake-equation/.

32. The Martian craze and the career of Percival Lowell form a significant chapter of the extraterrestrial life debate that I cannot address here.

33. The term, if not the idea itself, was coined by Robin Hanson in "The Great Filter—Are We Almost Past It?," 15 September 1998; http://mason.gmu.edu/~rhanson/greatfilter.html.

34. In his original formulation, Hanson proposes nine stages: The right star system (including organics); reproductive something (e.g., RNA); simple (prokaryotic) single-cell life; complex (archaeatic and eukaryotic) single-cell life; sexual reproduction; multi-cellular life; tool-using animals with big brains; where we are now; colonization explosion.

35. Avi Loeb, *Extraterrestrial: The First Sign of Intelligent Life Beyond Earth* (Boston: Houghton Mifflin Harcourt, 2021), 123.

36. They "wore tie clip-like dolphin insignia, and sent each other coded messages to test their readiness for extraterrestrial contact." D. Graham Burnett, *The Sounding*

of the Whale: Science and Cetaceans in the Twentieth Century (Chicago: University of Chicago Press, 2012), 612.

37. In one brief, hopeful moment in 1971, an engineer from Hewlett-Packard named Bernard Oliver initiated a large-scale search. His Project Cyclops failed to acquire significant NASA funding, certainly not the $10 billion that he requested, but it did result in a famous, 243-page report that has served as a blueprint for subsequent endeavors. See Lawrence Squeri, *Waiting for Contact: The Search for Extraterrestrial Intelligence* (Gainesville: University of Florida Press, 2016), 100–109.

38. Squeri, *Waiting for Contact*, 3.

39. See https://www.seti.org/faq. This institute has mostly been funded by private donors.

40. Tesla, "Talking with the Planets," *Collier's Weekly*, 9 February 1901, 4–5.

41. Daniel Chua and Alexander Rehding, *Alien Listening: Voyager's Golden Record and Music from Earth* (New York: Zone, 2021), 131.

42. Jonathan Sterne and Mitchell Akiyama, "The Recording That Never Wanted To be Heard and Other Stories of Sonification," *Oxford Handbook of Sound Studies*, ed. K. Bijsterveld and T. Pinch (New York: Oxford University Press, 2011), 544–60, 550.

43. Liu Cixin, *The Three-Body Problem*, translated by Ken Liu (New York: Tom Doherty Associates, 2008), 178.

44. Liu, *Three-Body Problem*, 271–72.

45. *Project Cyclops: A Design Study of a System for Detecting Extraterrestrial Intelligent Life* (Stanford/NASA/Ames Research Center, 1971), 170.

46. Horowitz, "We Didn't Find Anything, So the Search Goes On," (no date); https://radiocontact.org/portfolio/paulhorowitz/. I thank Ian Coss for bringing this interesting audio document to my attention.

47. Gary Tomlinson, *Metaphysical Song: An Essay on Opera* (Princeton, NJ: Princeton University Press, 1999).

48. Szendy, "Auditory Re-Turn (The Point of Listening)," in *Thresholds of Listening: Sound, Technics, Space*, ed. Sander van Maas (New York: Fordham University Press, 2015), 18–29, 25; Martin Heidegger, *The Principle of Reason*, trans. Reginald Lilly (Bloomington: Indiana University Press, 1991), 48.

49. This is a peculiar book co-authored across the Iron Curtain with Soviet scientist I. S. Shklovskii. The book was written originally by Shklovskii in Russian (most of it was published in 1963) and then published in English translation with additions by Sagan. The thought experiment discussed here is one of Sagan's additions. See I. S. Shklovskii and Carl Sagan, *Intelligent Life in the Universe*, trans. Paula Fern (San Francisco: Holden-Day, 1966).

50. Shklovskii and Sagan, *Intelligent Life in the Universe*, 439.

51. Norbert Wiener, *The Human Use of Human Beings: Cybernetics and Society* (London: Free Association Books, 1989 [1950]), 124.

52. John Durham Peters explains the logic of SETI: "To be taken as a message, a signal must have an extremely low probability of being either a random or a natural product. Russian exoscientists [in particular] made 'artificiality criteria' a topic of very sophisticated study, including analysis of the statistical properties of signals." See *Speaking into the Air: A History of the Idea of Communication* (Chicago: University of Chicago Press, 1999), 253.

53. Shklovskii and Sagan, *Intelligent Life in the Universe*, 423.

54. Shklovskii and Sagan, *Intelligent Life in the Universe*, 351.

55. Shklovskii and Sagan, *Intelligent Life in the Universe*, 351.

56. The ones and zeros, therefore, bear an iconic relationship to the imagined radio wave signal.

57. I am leaving out some subtleties of Sagan's argument here for the sake of simplicity.

58. Shklovskii and Sagan, *Intelligent Life in the Universe*, 425.

59. In addition to the obvious humanlike shape at the bottom of the image, Drake hoped the receiver would garner much additional information from this image. Can you guess what?

60. Steven J. Dick, *Life on Other Worlds: The 20th-Century Extraterrestrial Life Debate* (Cambridge: Cambridge University Press, 1998), 172.

61. Howard Smith, "Alone in the Universe," *Zygon* 51.2 (2016): 507. Published as part of a dossier titled "Exoplanets and Astrotheology."

62. Sara Seager, quoted in Smith, "Alone in the Universe," 507.

63. Helmreich, *Sounding the Limits of Life: Essays in the Anthropology of Biology and Beyond* (Princeton, NJ: Princeton University Press, 2016), 84. The emphasis is Helmreich's.

64. Andrew Lockley and Daniele Visioni, "Detection of Pre-Industrial Societies on Exoplanets," *International Journal of Astrobiology* 20 (2021): 73–80.

65. Matt Vella, "Read the Inspiring 'Questions of Existence' Letter from the World's Greatest Thinkers," *Time*, 20 July 2015; https://time.com/3964301/breakthrough-listen-letter/.

66. Vella, "Read the Inspiring 'Questions of Existence.'"

67. Chua and Rehding, *Alien Listening*, 168.

68. See the description of the group "Biomedical Engineering of Life" at the Max Planck Institute for Medical Research (November 25, 2019); https://www.mr.mpg.de/14414373/Biophysical-Engineering.

69. See, for example, Mike Wall, "We're Probably Living in a Simulation, Elon Musk Says," *Space.com* (7 September 2018); https://www.space.com/41749-elon-musk-living-in-simulation-rogan-podcast.html.

70. Nick Bostrom, "Are We Living in a Computer Simulation?" *Philosophical Quarterly* 53.211 (2003): 243.

71. Bostrom, "Computer Simulation?," 247–48; emphasis in original.

72. Marcelo Gleiser, "Why Reality Is Not a Video Game—And Why It Matters," *NPR*, 9 March 2017; https://www.npr.org/sections/13.7/2017/03/09/519376356/why-reality-is-not-a-video-game-and-why-it-matters.

73. Nick Bostrom, "Existential Risk: Analyzing Human Extinction Scenarios and Related Hazard," *Journal of Evolution and Technology* 9.1 (2002); http://jetpress.org/volume9/risks.html.

74. Srinivasan, "Stop the Robot Apocalypse," *London Review of Books* 37.18 (2015); https://www.lrb.co.uk/the-paper/v37/n18/amia-srinivasan/stop-the-robot-apocalypse. I thank Harvey Lederman for bringing this topic and this article to my attention.

75. Srinivasan, "Stop the Robot Apocalypse."

76. Achille Mbembe, "Theory from the Antipodes," *Johannesburg Salon* 5 (2012): 18–25, 25, emphasis in original.

77. "The Matrix Decoded: Le Nouvel Observateur Interview with Jean Baudril-

lard," trans. Gary Genosko and Adam Bryx (translation of an interview by Aude Lancelin for *Le Nouvel Observateur*, 19–25 June 2003), *International Journal of Baudrillard Studies* 1.2 (2004); https://baudrillardstudies.ubishops.ca/the-matrix-decoded-le-nouvel-observateur-interview-with-jean-baudrillard/.

78. Baudrillard, *Impossible Exchange*, trans. Chris Turner (London: Verso, 2001 [1999]), 3.

79. Plato, *Meno*, in *Plato's Complete* Works, ed. John M. Cooper (Indianapolis: Hackett, 1997), 886.

80. There is something patently absurd about Meno's paradox, as well as similar philosophical propositions such as solipsism. Quentin Meillassoux writes, "Schopenhauer said that solipsism was a fortress impossible to penetrate, but also pointless to attack, since it is empty. Solipsism is a philosophy nobody can refute, but also one that nobody can believe." See "Presentation by Quentin Meillassoux," in "Speculative Realism," *Collapse* 3: 307–450. Meillassoux's presentation appears on pp. 408–49. The passage quoted here appears on p. 423.

81. In the twenty-first century, Meno's paradox was rearticulated in the form of a dogma that Meillassoux terms "correlationism." See his *After Finitude: An Essay on the Necessity of Contingency*, trans. Ray Brassier (London: Continuum, 2008).

82. Baudrillard, *Impossible Exchange*, 151.

83. Hannah Arendt, *The Human Condition* (Chicago: University of Chicago Press, 1998), 320, 283.

84. I agree with Manuel DeLanda that no such enterprise exists: there is no such thing as "Science," only scientific domains. See his *Philosophical Chemistry: Genealogy of a Scientific Field* (New York: Continuum, 2015).

85. Climate change denialism might be on its last legs. Today, even the majority of conservatives accept the truth of climate change—but not at all in a good way. See Oliver Milman, "Climate Denial Is Waning on the Right: What's Replacing It Might Be Just as Scary," *The Guardian* 21 November 2021; https://www.theguardian.com/environment/2021/nov/21/climate-denial-far-right-immigration.

86. Sylvia Wynter, "Unsettling the Coloniality of Being/Power/Truth/Freedom Towards the Human, After Man, Its Overrepresentation—An Argument," *CR: The New Centennial Review* 3.3 (2003): 257–337.

Chapter 4

1. Alexis Pauline Gumbs, *Dub: Finding Ceremony* (Durham, NC: Duke University Press, 2020), 18.

2. Ronald Judy, "Kant and the Negro," *Surfaces* 1.8 (1991): 1–70.

3. Kant, quoted in Judy, "Kant and the Negro," 9.

4. Kant, quoted in Judy, "Kant and the Negro," 10.

5. What they show is that Kant's two fundamental "operations (synthesis of perceptions and attributions of meaning) are heterogenous to each other and only the fundamentally indeterminate designation/signification of language makes them coexist." This is why Blackness cannot ever be securely brought under the concept "Negro" with any operation—empirical, transcendental, or otherwise. The concept of the Negro, says Judy, short-circuits the entire point of the Kantian project: namely,

to establish limits for thought. "What the concept of The Negro exhibits," says Judy, "is that the very process of reflection on limits entails a labor that transcends them" (Judy, "Kant and the Negro," 71–72).

6. John Lilly, *Man and Dolphin: Adventures on a New Scientific Frontier* (New York: Doubleday, 1961), 217.

7. Daniel Chua and Alexander Rehding, *Alien Listening: Voyager's Golden Record and Music from Earth* (New York: Zone, 2021), 168.

8. Frantz Fanon, trans. Roland Judy and quoted in Judy, "Kant and the Negro," 7.

9. Judy, "Kant and the Negro," 7.

10. Nahum Dmitri Chandler, *X: The Problem of the Negro as a Problem for Thought* (New York: Fordham University Press, 2014).

11. Jared Sexton, "The Social Life of Social Death," *InTensions*, 5 (Fall/Winter 2011): 1–47, 6–7

12. Kodwo Eshun, *More Brilliant Than the Sun: Adventures in Sonic Fiction* (London: Quartet, 1998), 00[-006]-00[-005]; Eshun's pagination.

13. Sun Ra, *Space Is the Place* (film; dir. John Coney, 1974).

14. Dumas and Sun Ra knew each other and collaborated briefly in 1966.

15. See, for example, Rachel Poser's article on Dan-El Padilla-Peralta, "He Wants to Save Classics from Whiteness: Can the Field Survive?," *New York Times*, 2 February 2021; https://www.nytimes.com/2021/02/02/magazine/classics-greece-rome-whiteness.html.

16. Szwed, *Space Is the Place: The Lives and Times of Sun Ra* (Durham, NC: Duke University Press, 2020 [1997]), 108, 130.

17. Szwed, *Space Is the Place*, 21; my emphasis. At the Booker T. Washington Branch of the public library, he was also exposed to "revisionist histories of eighteenth-century Prince Hall Freemasonry, in which the ancient cities of Ethiopian peoples assume a central role in early civilization and black masons are present at the founding of the American republic." William Sites, *Sun Ra's Chicago: Afrofuturism and the City* (Chicago: University of Chicago Press, 2020), 21.

18. Szwed, *Space Is the Place*, 132.

19. Sun Ra, quoted in Szwed, *Space Is the Place*, 131. Throughout this chapter, when Szwed is quoting Sun Ra from another source, I cite only Szwed without providing the original source. I do this, first, because the full citation information is not particularly relevant, and, second, because the citation information is sometimes confusing. Rather than risk providing an inaccurate citation, I cite Szwed's book, which the interested reader may consult for further details. Since I quote so often from Szwed's book in the rest of this section on Sun Ra, subsequent citations are given in the text by page no. in parentheses.

20. One may ask, "What about a statement like, 'Your hair is on fire'?" This is a deceptively good example to prove the point. Why? Of course, the truth of the warning about one's hair being on fire is important, but just as important is the capacity to judge the statement's relevance. If someone says, "Your hair is on fire," and another person says, "You have nice shoes," it's just as important to know which of these statements is relevant as it is to know which is true. This means that importance and relevance are just as valuable as truth. For a philosophical argument in this direction, see Manuel DeLanda, *Intensive Science and Virtual Philosophy* (London: Continuum, 2002).

21. Sun Ra, quoted in Sites, *Sun Ra's Chicago*, 11.

22. Sites, *Sun Ra's Chicago*, 11.

23. Amiri Barka, quoted in Brent Hayes Edwards, "The Race for Space: Sun Ra's Poetry," in *Sun Ra: The Immeasurable Equation: The Collected Poetry and Prose*, ed. James L. Wolf and Hartmut Geerkun (Warteweil, Germany: Waitawhile, 2005), 29–55, 30.

24. See "Kapwani Kiwanga in Conversation with Brienne Walsh," *Ocula*, 30 March 2016; https://ocula.com/magazine/conversations/kapwani-kiwanga/.

25. Although she is fascinated by Afrofuturism as "an 'object' of study," she is hesitant to identify too closely with the term, in part because it is so "slipperily defined." Personal correspondence, 27 June 2016.

26. Kiwanga describes the series as follows: "The performance creates a space in which fiction and historical facts converge. Eclectic archives are drawn from specialist and popular sources. The fictional elements of the performance centre on *The United States of Africa Space Agency*, a fabricated pan-African organisation established in the year 2058." This description is from a document, simply called "Afrogalactica," shared privately with me in 2015.

27. Unless stated otherwise, all further quotes from Kiwanga are from *The Deep Space Scrolls*.

28. As part of the festival, events were held simultaneously in Johannesburg (South Africa), Lagos (Nigeria), and Nairobi (Kenya) in October 2015.

29. For more on what Kiwanga does with Zimbabwe, see Gavin Steingo, "Kapwani Kiwanga's Alien Speculations," *Images Re-vues: Histoire, anthropologie et théorie de l'art* 14 (2017); http://imagesrevues.revues.org/4083.

30. Although there are in fact two kinds of supernovas, Kiwanga is interested primarily in those that take place at the end of a star's life cycle. On the two kinds of supernovas, see http://www.nasa.gov/audience/forstudents/5–8/features/nasa-knows/what-is-a-supernova.html.

31. Mary Douglas, "If the Dogon," *Cahiers d'Études Africaines* 7.28 (1967): 659–72, 659. See also Marcel Griaule, *Dieu d'Eau: Entretiens avec Ogotemmêli* (Paris: Éditions du Chêne, 1948), published in English as *Conversations with Ogotemmêli* (London: Oxford University Press, 1965); Marcel Griaule and Germaine Dieterlen, *Le Renard Pâle* (Paris: Institut d'ethnologie, 1965), published in English as *The Pale Fox*, trans. Stephen C. Infantino (Chino Valley, AZ: Continuum Foundation/Afrikan World Books, 1986).

32. Douglas, "If the Dogon," 660.

33. Douglas, "If the Dogon," 660.

34. See Griaule, *Conversations with Ogotemmêli*: "God [Amma] had further intercourse with his earth-wife. . . . Two beings were thus formed" (18).

35. Genevieve Calame-Griaule, quoted in Douglas, "If the Dogon," 664. Like her father, Marcel Griaule, Calame-Griaule was a well-known ethnographer of the Dogon people. See *Ethnologie et langage: La parole chez les Dogon* (Paris: Gallimard, 1965), 546–47.

36. Douglas, "If the Dogon," 667. Cf. Griaule and Dieterlen, *The Pale Fox*, 240, 287–89.

37. Douglas, "If the Dogon," 664.

38. Marcel Griaule and Germaine Dieterlen, "Un système soudanais de Sirius," *Journal de la Société des Africanistes* 20 (1950): 273–94, 274. My translation. The original reads, "La question n'a pas été tranchée, ni même posée, de savoir comment des hommes ne disposant d'aucun instrument connaissent les mouvements et certaines caractéristiques d'astres pratiquement invisibles."

39. Robert Temple, *The Sirius Mystery* (Rochester, VT: Destiny Books, 1998 [1976]).

40. Erich von Däniken, *Erinnerungen an die Zukunft: Ungelöste Rätsel der Vergangenheit* (translated into English as *Chariots of the Gods? Unsolved Mysteries of the Past*, published in 1969); Peter Kolosimo, *Not of This World*, trans. A. D. Hills (New Hyde Park, NY: University Books, 1971).

41. Karl Mauch, *Journals of Carl Mauch: His Travels in the Transvaal and Rhodesia, 1869–1872*, transcribed from the original by Mrs. E. Bernhard, translated by F. O. Bernhard, edited by E. E. Burke (Salisbury: National Archives of Rhodesia, 1969); James Theodore Bent, *The Ruined Cities of Mashonaland: Being a Record of Excavation and Exploration in 1891*, 2nd ed. (London: Longmans, Green, 1893); on Breuil, see Alan Houghton Brodick, *Father of Prehistory: The Abbé Henri Breuil: His Life and Times* (New York: H. Wolff, 1963). See also Steingo, "Kapwani Kiwanga's Alien Speculations."

42. Walter E. A. van Beek, "Dogon Restudied: A Field Evaluation of the Work of Marcel Griaule," *Current Anthropology* 32.2 (1991): 139–58; and "Haunting Griaule: Experiences from the Restudy of the Dogon," *History in Africa* 31(2004): 43–68.

43. The reference here is likely to the *Nommo anagonno*, the first living beings from whom humankind is thought to have descended. See *The Pale Fox*, 157.

44. Fanny Curtat, "The 'Undecidable' Fetish at Work: Benoit Pype and Kapwani Kiwanga," trans. Bernard Schütze, *Espace* 113 (2016): 37–43.

45. See Jocelyne Guilbault and Roy Cape, *Roy Cape: A Life on the Calypso and Soca Bandstand* (Durham, NC: Duke University Press, 2014); Alejandro Madrid, *In Search of Julián Carrillo and Sonido 13* (Oxford: Oxford University Press, 2015).

46. Douglas, "If the Dogon," 667. Cf. Griaule and Dieterlen, *The Pale Fox*, 240, 287–89.

47. I am grateful to Brent Hayes Edwards for sharing information with me about this topic over the years.

48. Simone White, *Dear Angel of Death* (Brooklyn: Ugly Duckling, 2018), 77. See also, Amiri Baraka, *Digging: The Afro-American Soul of American Classical Music* (Berkeley: University of California Press, 2009).

49. White, *Dear Angel of Death*, 79.

50. White, *Dear Angel of Death*, 82.

51. White, *Dear Angel of Death*, 86.

52. The term "social death" is from Orlando Patterson, *Slavery and Social Death* (Cambridge, MA: Harvard University Press, 1982). Sexton and Wilderson adjust Patterson's definition of the term.

53. Nicole Mitchell, "What Was Feared Lost," in *Arcana VIII: Musicians on Music*, ed. John Zorn (New York: Hips Road, 2017), 221–36, 222.

54. Mitchell, "What Was Feared Lost," 222.

55. Mitchell, "What Was Feared Lost," 223, emphasis added.

56. Mitchell, "What Was Feared Lost," 234, emphasis in original.

57. Mitchell, "What Was Feared Lost," 234.

58. In summarizing Sun Ra's core ideas, especially as he taught them at Berkeley in 1971, Szwed offers the following: "Blacks were originally in direct contact with God, and the other races had to reach Him through them. But black people did something wrong, lost contact, fell, for spiritual reasons, leaving them with little but music, a gift of the Creator" (Szwed, *Space Is the Place*, 310).

59. Szwed, *Space Is the Place*, 213.

60. Mackey, "Sound and Sentiment, Sound and Symbol," in *Discrepant Engagement: Dissonance, Cross-Culturality, and Experimental Writing* (Cambridge: Cambridge University Press, 1993), 259.

61. White, *Dear Angel of Death*, 95.

62. White, *Dear Angel of Death*, 101.

63. On natal alienation, see Patterson, *Slavery and Social Death*; cf. Frank B. Wilderson III, *Red, White & Black: Cinema and the Structure of US Antagonisms* (Durham, NC: Duke University Press, 2010).

64. Wilderson, 279; emphasis in original. For Wilderson, the Middle Passage, rather than Africa, is the (non-)origin of Blackness. I return to this briefly in the epilogue.

65. When used for a branch of contemporary Black critical theory, Afro-pessimism means something distinct from the other meaning of the term—namely, that Africa as a continent is without hope due to failed states, poverty, and so on.

66. It's interesting that Mitchell has nothing to say about The Stone as an actual space of musical performance (even though she is very familiar with it). What I and many others found so compelling about The Stone was its almost aggressive asceticism: it was utterly functionalist as a venue, with no design frills, no bar, and no atmosphere. The Stone was, for all intents and purposes, a dry storefront recital hall that was all about the music. It could probably be said that the venue's metaphysical aspirations were undergirded by an outright rejection of extravagance. (I speak about The Stone in the past tense because, even though it technically still exists, it is a fundamentally different venue since its move to The New School in 2018.)

67. More pointedly (and more controversially), I would argue that Mitchell's subject position does compel critical reflection in a way that Lilly's did not. As Fumi Okiji writes, "Black life cannot help but be lived as critical reflection. One need not be politically committed to question the integrity of the world. Blackness is a mode of existence in which the disjuncture between the reality of one's everyday living and the ways one is understood by society at large is so pronounced that the former must be considered an impossibility or lie in order to preserve the latter." *Jazz as Critique: Adorno and Black Expression Revisited* (Stanford, CA: Stanford University Press, 2018), 5.

68. White, *Dear Angel of Death*, 118; emphasis in original.

69. White, *Dear Angel of Death*, 120.

70. White, *Dear Angel of Death*, 125. The phrase "constant improvisational contact" is from Fred Moten, "Blackness and Nothingness (Mysticism in the Flesh)," *South Atlantic Quarterly* 112.4 (2013): 737–80, 754.

71. "Whatever you want. I serve that, for money, for a nihilistic, endlessly repetitive and narcotized kind of peace" (White, *Dear Angel of Death*, 149).

72. White, *Dear Angel of Death*, 148–49; emphasis in original. On page 147, White had quoted a devastating poem by Emily Dickinson about a "pain . . . so utter" it "Would drop him—Bone by Bone."

73. Wilderson, *Red, White & Black*, 37.

74. It is worth noting that while many have reacted aggressively to the claim that anti-Blackness constitutes *the world*, many White philosophers have made similarly strong claims with hardly a peep of retort. The French philosopher Alain Badiou, for example, has been arguing for decades that civil society constructs a world and that only a singular revolutionary "Event" can usher in a new one. For Badiou, the

dominant structure of our world is capitalism. Interestingly, however, he seems reluctant to systematically theorize capitalism qua world. See Nick Nesbitt, "Capital, Logic of the World," in "The Concept of World in Contemporary Philosophy," ed. Rok Benčin, special issue, *Filozofski vestnik* 42.2 (2021): 159–91.

75. The key text here is Bernard Le Bovier de Fontenelle's *Entretiens sur la pluralité des mondes* (1686).

76. More precisely, *der Stein is weltlos, das Tier is weltarm, der Mensch ist weltbildend* (the stone is world-less, the animal is world-poor, the human is world-forming). Martin Heidegger, *Die Grundbegriffe der Metaphysik*, vols. 29/30 of *Gesamtausgabe* (Frankfurt am Main: Vittorio Klostermann, 1983 [1938]), 261.

77. Georg Wilhelm Friedrich Hegel, *The Philosophy of History* (New York: Dover, 1956 [1837]), 99.

78. "The Year Afropessimism Hit the Streets? A Conversation at the Edge of the World: Aaron Robertson Talks to Frank Wilderson III," *Literary Hub*, 27 August 2020; https://lithub.com/the-year-afropessimism-hit-the-streets-a-conversation-at-the-edge-of-the-world/.

79. Kara Keeling, *Queer Times, Black Futures* (New York: New York University Press, 2018), 67.

80. Keeling, *Queer Times*, 67.

81. Wilderson, *Red, White & Black*, 96.

82. Wilderson, *Red, White & Black*, 281; emphasis in original.

83. Ralph Ellison, *Invisible Man* (New York: Vintage Books, 1995 [1952]), 508.

84. "Acknowledgments," in Marcel Griaule and Germaine Dieterlen, *The Pale Fox*, trans. Stephen C. Infantino (Chino Valley, AZ: Continuum Foundation/Afrikan World Books, 1986), 17. See also Ida P. Moffett's "Introduction to the English Translation" in the same edition.

85. Achille Mbembe, *Critique of Black Reason*, trans. Laurent Dubois (Durham, NC: Duke University Press, 2017), 179.

86. Mbembe, *Critique of Black Reason*, 4.

87. Mbembe, *Critique of Black Reason*, 179.

88. Mbembe, *Critique of Black Reason*, 181–82.

89. I thank Walter van Beek for providing me with this reference.

90. Mike Sell, *Avant-Garde Performance and the Limits of Criticism* (Ann Arbor: University of Michigan Press, 2005).

91. About Allan Kaprow's *Eighteen Happenings in Six Parts*, Sell says: "Kaprow's viewer-spectator is displaced from spectatorial authority—he or she can't tell where the boundaries of the event are to be found." See Sell, *Avant-Garde Performance*, 144.

92. Mbembe, *Critique of Black Reason*, 38.

93. Mbembe, *Critique of Black Reason*, 28.

94. Michael Schetsche, René Gründer, Gerhard Mayer, and Ina Schmied-Knittel, "Der maximal Fremde: Überlegungen ze einer transhumanen Handlungstheorie," *Berliner Journal für Soziologie* 3 (2009): 469–91.

95. Szwed, *Space Is the Place*, 383.

96. Sites, *Sun Ra's Chicago*, 43.

97. As quoted in Tam Tiofori, "Sun Ra's Space Odyssey," *Down Beat*, 14 May 1970, 14.

98. Jean-François Lyotard, *The Postmodern Explained: Correspondence 1982–1985* (Minneapolis: University of Minnesota Press, 1992).

99. Okiji, *Jazz as Critique*, 53. Here, Okiji is speaking specifically about Fred Moten's approach.

100. Sexton, "Social Life of Social Death," 28.

101. That something, says Sexton, is how Black "life is *lived* in social *death*. Double emphasis, on lived and on death." Later in the same essay, he writes, "Blackness is not the pathogen in afro-pessimism, the world is. Not the earth, but the world" (Sexton, "Social Life of Social Death," 29, 31).

Epilogue

1. Although allusions are plentiful throughout this book, I have deliberately avoided lengthy theoretical digressions into theories of autoimmunity à la Derrida, Esposito, and so on, in order not to overburden the text with theoretical jargon.

2. Kodwo Eshun, *More Brilliant Than the Sun: Adventures in Sonic Fiction* (London: Quartet, 1998), 00[-006]-00[-005]; Eshun's pagination.

3. Wilderson adjusts Orlando Patterson's use of the term in *Slavery and Social Death* (Cambridge, MA: Harvard University Press, 1982).

4. "Introduction to *The Social Life of Social Death*: Julian Warner in Conversation with Frank B. Wilderson III," 27 February 2022; https://www.kurzfilmtage.de/en/channel/the-social-life-of-social-death/introduction-to-the-social-life-of-social-death-julian-warner-in-conversation-with-frank-b-wilderson-iii/.

5. Szwed, *Space Is the Place: The Lives and Times of Sun Ra* (Durham, NC: Duke University Press, 2020 [1997]), 140.

6. Aaron Robertson, "The Year Afropessimism Hit the Streets? A Conversation at the Edge of the World: Aaron Robertson Talks to Frank Wilderson III," *Literary Hub*, 27 August 2020; https://lithub.com/the-year-afropessimism-hit-the-streets-a-conversation-at-the-edge-of-the-world/.

7. Grebowicz, *Whale Song* (New York: Bloomsbury, 2017), 111; Joanna Bourke, *Loving Animals: On Bestiality, Zoophilia and Posthuman Love* (London: Reaktion, 2020). Bourke adapts the theory of "parental embodiment mentalizing" from Dana Shai and Jay Belsky for human-animal relations (Bourke, *Loving Animals*, 127–30).

8. Amia Srinivasan, "What Does Fluffy Think? (Review of *Loving Animals: On Bestiality, Zoophilia and Posthuman Love*, by Joanna Bourke)," *London Review of Books* 43.19 (2021); https://www.lrb.co.uk/the-paper/v43/n19/amia-srinivasan/what-does-fluffy-think.

9. Terrence Deacon, *The Symbolic Species: The Co-evolution of Language and the Brain* (New York: W. W. Norton, 1997), 22.

10. Georg Wilhelm Friedrich Hegel, *The Philosophy of History* (New York: Dover, 1956 [1837]), 99.

11. On some historical forms of slavery, see Igor Kopytoff, "The Cultural Biography of Things: Commoditization as Process," in *The Social Life of Things: Commodities in Cultural Perspective*, ed. Arjun Appadurai (Cambridge: Cambridge University Press, 1986), 64–91. On Blackness and slavery, see Wilderson, *Red, White & Black*, especially chap. 1.

12. "I must begin. Begin with what? The only thing in the world that's worth beginning: The End of the World, no less." Aimé Césaire, *Return to My Native Land* (New York: Archipelago Books, 1969 [1956]), 39.

13. Gregory Bateson, *Steps to an Ecology of Mind* (Chicago: University of Chicago Press, 1999 [1972]), 501.

14. In 2018, the Tiktaalik (the evolutionary link between fish and land-dwelling vertebrates) became the basis of a series of misanthropic memes suggesting, often humorlessly, that the creature should have stayed under water.

15. Website of the Voluntary Human Extinction Movement; http://www.vhemt.org/.

16. Andrew Gibson, *Misanthropy: The Critique of Humanity* (London: Bloomsbury, 2017).

17. Molière, quoted in Gibson, *Misanthropy*, 2.

18. I thank Kostis Velonis for this insight.

19. Gibson, *Misanthropy*, 3.

20. Gibson seems to recognize this rupture, if only implicitly, in the concluding chapter of his book.

21. Herbert S. Terrace, "Evidence for Sign Language in Apes: What the Ape Signed or How Well the Ape Was Loved?" *Contemporary Psychology* 27.1 (1982): 67–68. This short essay is a response to a review of his own book. See Beatrix T. Gardner, "Project Nim: Who Taught Whom?" (review of the book *Nim: A Chimpanzee Who Learned Sign Language*, by Herbert S. Terrace [1979])," *Contemporary Psychology* 26.6 (1981): 425–26. For other relevant work in this area, see Herbert S. Terrace, *Nim: A Chimpanzee Who Learned Sign Language* (New York: Knopf, 1979); Allen Gardner, Beatrix T. Gardner, and Thomas E. Van Cantfort, eds., *Teaching Sign Language to Chimpanzees* (Albany: State University of New York Press, 1989); Herbert S. Terrace, *Why Chimpanzees Can't Use Language and Only Humans Can* (New York: Columbia University Press, 2019).

22. Gardner, "Project Nim: Who Taught Whom," 426. The social basis of language acquisition is so important to the Gardners that they state it explicitly on the first page of the preface and the first page of the introduction to their book. See R. Allen Gardner and Beatrix T. Gardner, "Preface," and "Cross-fostering Laboratory," in A. Gardner, B. Gardner and Van Cantfort, *Teaching Sign Language to Chimpanzees*.

23. All quotations in this paragraph are from the documentary *Project Nim* (dir. James Marsh), 2011.

24. Bill Tynan (a Project Nim teacher) referred to the room in question as a "dungeon of a classroom" (*Project Nim*, 2011).

25. Terrace, "Evidence for Sign Language in Apes."

26. See Oskar Pfungst, *Das Pferd der Herrn von Osten: Der Kluge Hans* (Leipzig: Verlag von Johann Ambrosius Barth, 1907).

27. Among them were Irene Pepperberg, Duane Rumbaugh, Sue Savage, and Diana Reiss. See Irene Pepperberg, *Alex & Me: How a Scientist and a Parrot Discovered a Hidden World of Animal Intelligence—and Formed a Deep Bond in the Process* (New York: HarperLuxe, 2008).

28. This is James Johnson's description based on contemporary reports and not a contemporary report itself (*Listening in Paris* [Berkeley: University of California Press, 1994], 130).

29. See, for example, Edward Helmore, "Google Engineer Says AI Bot Wants to 'Serve Humanity' but Experts Dismissive," *The Guardian* 12 June 2022; https://www.theguardian.com/technology/2022/jun/13/google-ai-bot-sentience-experts-dismissive-blake-lemoine.

30. Ian Bogost, "Google's 'Sentient' Chatbot Is Our Self-Deceiving Future,"

Atlantic, 14 June 2022; https://www.theatlantic.com/technology/archive/2022/06/google-engineer-sentient-ai-chatbot/661273/.

31. Weizenbaum, quoted in Bogost, 2022. Weizenbaum's *Computer Reason and Human Reason* was first published in 1976 by W. H. Freeman.

32. John Durham Peters, *Speaking into the Air: A History of the Idea of Communication* (Chicago: University of Chicago Press, 1999), 237.

33. The Mechanical Turk was an invention of Wolfgang von Kempelen for the Empress Maria Theresa of Austria.

34. See Louise Chude-Sokei, *The Sound of Culture: Diaspora and Black Technopoetics* (Middletown, CT: Wesleyan University Press, 2015), 22. .

35. Chude-Sokei, *Sound of Culture*, 22.

36. P. T. Barnum, *Struggles and Triumphs; or, Forty Years' Recollections of P. T. Barnum, Written by Himself* (Buffalo, NY: Warren, Johnson & Co., 1873), 157.

37. Chude-Sokei, *Sound of Culture*, 24.

38. P. T. Barnum, *Struggles and Triumphs*, 157; Chude-Sokei, *Sound of Culture*, 24.

39. Ralph Ellison, *Invisible Man* (New York: Vintage Books, 1995), 3; emphasis in original.

40. David Rothenberg, *Thousand Mile Song: Whale Music in a Sea of Sound* (New York: Basic Books, 2008), 212. Elsewhere, Rothenberg seems more sure of the connection. See, for example, his "Interspecies Improvisation," in *Oxford Handbook of Critical Improvisation Studies*, vol. 1, ed. George E. Lewis and Benjamin Piekut (Oxford: Oxford University Press, 2016), 500–22.

41. Steven J. Dick, *Astrobiology, Discovery, and Societal Impact* (Cambridge: Cambridge University Press, 2018).

42. Let's be honest: if an alien civilization manages to land on planet Earth, any army we send won't last a minute.

43. Ed Yong, *An Immense World: How Animal Senses Reveal the Hidden Realms Around Us* (New York: Random House, 2022), 5. Yong makes his debt to Jakob von Uexküll and the notion of *Umwelt* explicit.

44. Eduardo Kohn, *How Forests Think: Toward an Anthropology Beyond the Human* (Berkeley: University of California Press, 2013).

Appendix

1. I am leery of biosemioticians who espouse what Gary Tomlinson calls (critically) "semantic universalism." For an excellent discussion, see Tomlinson, *The Machines of Evolution and the Scope of Meaning* (New York: Zone, 2023), 35–37.

2. Here, I work with the roughly standardized schema adopted by most cotemporary authors. But Peirce himself changed the terms of his typology throughout his life. For a good start, see Nathan Houser and Christian Kloesel, eds., *The Essential Peirce: Selected Philosophical Writings*, vol. 1 (1867–1893) (Bloomington: Indiana University Press, 1992). Page numbers provided in these notes are for this edition.

3. See Gary Tomlinson, *A Million Years of Music: The Emergence of Human Modernity* (New York: Zone Books, 2015), 189.

4. Tomlinson writes, "There is in the interpretant no interpretation at all, in the usual sense of this word. Instead, the interpretant is a calling of the organism by the sign vehicle that shapes the vehicle's mutually constrained relation to the object" (*Machines of Evolution*, 62).

5. But note the caveat: although controversial, some claim that several species of non-human animals can be trained by humans to use rudimentary symbolic sign systems.

6. "Indexical association relies on a certain amount of learning. . . . Associating the smell of smoke and the likelihood of a fire is a learned association, arising from past experiences where the smell of smoke did indeed accompany a fire." See Tanya de Villiers, "Why Peirce Matters: The Symbol in Deacon's *Symbolic Species*," *Language Sciences* 29 (2007): 88–108, 98.

7. Peirce defined the icon as "a sign which stands for something merely because it resembles it." "Icons are so completely substituted for their objects," he says, "as hardly to be distinguished from them." See "On the Algebra of Logic: A Contribution to the Philosophy of Notation" (1885), in Houser and Cloesel, eds., *The Essential Peirce*, 1:226.

8. Tomlinson, *Million Years of Music*, 190.

9. Gary Tomlinson, *Culture and the Course of Human Evolution* (Chicago: University of Chicago Press, 2018), 60.

10. See De Villiers, "Why Peirce Matters," 98.

11. Tomlinson, *Million Years of Music*, 190–91.

12. "As associative indexes multiply and accumulate through learning, there occurs something like a phase transition in cognition" (Tomlinson, *Million Years of Music*, 191). Tomlinson emphasizes that this new capacity is afforded by increased memory resources.

13. On emergence, see Manuel DeLanda, *Philosophy and Simulation: The Emergence of Synthetic Reason* (London: Continuum, 2011).

14. Eduardo Kohn, *How Forests Think: Toward an Anthropology Beyond the Human* (Berkeley: University of California Press, 2013), 38. His use of the term "provincializing" is a reference to Dipesh Chakrabarty, *Provincializing Europe: Postcolonial Thought and Historical Difference* (Princeton, NJ: Princeton University Press, 2000).

15. Kohn, *How Forests Think*, 41.

16. Kohn, *How Forests Think*, 27.

17. Kohn, *How Forests Think*, 29.

18. Kohn, *How Forests Think*, chap. 1, esp. 29–33.

19. David Rothenberg, *Thousand Mile Song: Whale Music in a Sea of Sound* (New York: Basic Books, 2008), 212; my emphasis.

20. Rothenberg, *Thousand Mile Song*, 212.

21. See Charles Darwin, *The Descent of Man, and Selection in Relation to Sex* (Princeton, NJ: Princeton University Press, 1981 [1871]), pt. 1, 56. See Steven Mithen, *The Singing Neanderthals: The Origins of Music, Language, Mind, and Body* (Cambridge, MA: Harvard University Press, 2007). For a critique of this aspect of Darwin's and Mithen's work, see Tomlinson, *Million Years of Music* (112–19). Sagan was less consistent in this regard, but see, for example, his *Murmurs of the Earth: The Voyager Interstellar Record* (New York: Random House, 1978), 13.

22. Tomlinson, *Million Years of Music*, 277

23. Tomlinson, *Million Years of Music*, 110–11. I do not mean to suggest that this terrain is simple or anything like unanimously agreed upon.

24. Tomlinson, *Million Years of Music*, 110.

25. Tomlinson, *Million Years of Music*, 203.

26. Tomlinson, *Million Years of Music*, 203; emphases in original.

27. Research on the topic reaches in wildly different directions. Of possible relevance is the stabilization of musical reference through so-called "topics" during the long eighteenth century (see, for example, *The Oxford Handbook of Topic Theory*, ed. Danuta Mirka [New York: Oxford University Press, 2016]), and the possibility of limited, culturally (perhaps even "universally"—although research seems to suggest otherwise) shared narrative projections onto music. See, for example, Elizabeth H. Margulis et al., "Narratives in Response to Instrumental Music Reveal Culture-Bounded Intersubjectivity," *Proceedings of the National Academy of Science* 119.4 (2021): e2110406119.

28. Agawu, *Music as Discourse: Semiotic Adventures in Romantic Music* (Oxford: Oxford University Press, 2000), 26.

29. Steven Feld and Aaron Fox, "Music and Language," *Annual Review of Anthropology* 23 (1994): 27; as cited in Agawu, *Music as Discourse*, 22.

30. Tomlinson, *Million Years of Music*, 205.

31. Manuel DeLanda, "The Virtual Breeding of Sound," in *Sound Unbound: Sampling Digital Music and Culture*, ed. Paul D. Miller (Cambridge, MA: MIT Press, 2008), 219. Tomlinson calls these offshoots "epicycles."

32. Tomlinson, *Million Years of Music*, 191.

33. Terrence Deacon, "The Symbol Concept," in *Oxford Handbook of Language Evolution*, ed. Maggie Tallerman and Kathleen R. Gibson (Oxford: Oxford University Press, 2012), 401.

34. In a similar manner, Tecumseh Fitch characterizes human music as having "discrete time and discrete pitch." See "The Biology and Evolution of Music: A Comparative Perspective," *Cognition* 100 (2006): 173–215, 179.

35. Tomlinson, *Million Years of Music*, 272.

36. In my book *Kwaito's Promise: Music and the Aesthetics of Freedom in South Africa* (Chicago: University of Chicago Press, 2016), I draw on Jacques Rancière's insight that if music is an illusion, it is only so in the sense that it *doubles* reality rather than masking it.

37. On music's timbral variety in relation to language, see Tomlinson, *Million Years of Music*, 278.

38. Rancière describes this ambiguity as part of "aesthetic art" as a whole. Here, I adapt it to describe music in particular. See Jacques Rancière, "The Aesthetic Revolution and Its Outcomes," *New Left Review* 14 (March/April 2002): 133–51, 151.

39. On the second point, the very notion of bird and whale *song* is confounding, since musicologists would define "song" as music sung with words—and none of the hundreds of bird or whale song researchers believe that the animals they study use words. (I thank Viktoria Tkaczyk for bringing this point to my attention.) In the course of writing this book, I was surprised to learn that non-specialists sometimes use the word "song" colloquially in a different way from musicologists. Several friends I asked defined song as a short piece or melody (even without words), a definition closer to Payne's and McVay's (neither of whom have any musical training). I address these thorny definitional issues in chapter 1.

40. I briefly mentioned African talking drums—drums that can and do convey symbolic meaning. Agawu seems to agree. See Agawu, *Music as Discourse*, 17. Elsewhere, Agawu notes that, because it is a tone language, Ewe is (always already) "on

its way to music." However, he says, the gap between language and music is "not ultimately eliminable." (The transition from language to music proper is breeched, at least in part, through "explicit, usually externalized meter.") "The interesting thing about the language-as-music and music-as-language metaphors," Agawu writes, "is to be found in this dynamic and unstable condition." See *African Rhythm: A Northern Ewe Perspective* (Cambridge: Cambridge University Press, 1995), 181.

41. Michael Gallope, *Deep Refrains: Music, Philosophy, and the Ineffable* (Chicago: University of Chicago Press, 2017).

42. Kohn, *How Forests Think*, 89. This is a response to Thomas Nagel's famous essay, "What Is It Like to Be a Bat?" *Philosophical Review* 83.4 (1974): 435–50.

43. The quote is from a review of Josiah Royce's *The Religious Aspect of Philosophy* prepared for the *Popular Science Monthly* but not published during Peirce's life. The review was later printed in Houser and Kloesel, eds., *The Essential Peirce*, 1:229–41.

Index